Martin Seyr

## Autonomous Mobile Robot Motion Control

Martin Seyr

# Autonomous Mobile Robot Motion Control

A concept for tracking control, navigation and motion planning

Südwestdeutscher Verlag für Hochschulschriften

**Impressum/Imprint (nur für Deutschland/ only for Germany)**
Bibliografische Information der Deutschen Nationalbibliothek: Die Deutsche Nationalbibliothek verzeichnet diese Publikation in der Deutschen Nationalbibliografie; detaillierte bibliografische Daten sind im Internet über http://dnb.d-nb.de abrufbar.
Alle in diesem Buch genannten Marken und Produktnamen unterliegen warenzeichen-, marken- oder patentrechtlichem Schutz bzw. sind Warenzeichen oder eingetragene Warenzeichen der jeweiligen Inhaber. Die Wiedergabe von Marken, Produktnamen, Gebrauchsnamen, Handelsnamen, Warenbezeichnungen u.s.w. in diesem Werk berechtigt auch ohne besondere Kennzeichnung nicht zu der Annahme, dass solche Namen im Sinne der Warenzeichen- und Markenschutzgesetzgebung als frei zu betrachten wären und daher von jedermann benutzt werden dürften.

Verlag: Südwestdeutscher Verlag für Hochschulschriften Aktiengesellschaft & Co. KG
Dudweiler Landstr. 99, 66123 Saarbrücken, Deutschland
Telefon +49 681 37 20 271-1, Telefax +49 681 37 20 271-0, Email: info@svh-verlag.de
Zugl.: Wien, TU, Diss., 2006

Herstellung in Deutschland:
Schaltungsdienst Lange o.H.G., Berlin
Books on Demand GmbH, Norderstedt
Reha GmbH, Saarbrücken
Amazon Distribution GmbH, Leipzig
ISBN: 978-3-8381-0632-8

**Imprint (only for USA, GB)**
Bibliographic information published by the Deutsche Nationalbibliothek: The Deutsche Nationalbibliothek lists this publication in the Deutsche Nationalbibliografie; detailed bibliographic data are available in the Internet at http://dnb.d-nb.de.
Any brand names and product names mentioned in this book are subject to trademark, brand or patent protection and are trademarks or registered trademarks of their respective holders. The use of brand names, product names, common names, trade names, product descriptions etc. even without a particular marking in this works is in no way to be construed to mean that such names may be regarded as unrestricted in respect of trademark and brand protection legislation and could thus be used by anyone.

Publisher:
Südwestdeutscher Verlag für Hochschulschriften Aktiengesellschaft & Co. KG
Dudweiler Landstr. 99, 66123 Saarbrücken, Germany
Phone +49 681 37 20 271-1, Fax +49 681 37 20 271-0, Email: info@svh-verlag.de

Copyright © 2009 by the author and Südwestdeutscher Verlag für Hochschulschriften Aktiengesellschaft & Co. KG and licensors
All rights reserved. Saarbrücken 2009

Printed in the U.S.A.
Printed in the U.K. by (see last page)
ISBN: 978-3-8381-0632-8

# Contents

**1 Introduction**     3
    1.1 Motivation . . . . . . . . . . . . . . . . . . . . . . . . . . . . . 3
    1.2 Scope . . . . . . . . . . . . . . . . . . . . . . . . . . . . . . . . 3

**2 Modelling**     5
    2.1 Introduction . . . . . . . . . . . . . . . . . . . . . . . . . . . . 5
    2.2 Mathematical description . . . . . . . . . . . . . . . . . . . . . 6
    2.3 Numerical simulation . . . . . . . . . . . . . . . . . . . . . . . 15
    2.4 Measurements . . . . . . . . . . . . . . . . . . . . . . . . . . . 16
    2.5 Validation . . . . . . . . . . . . . . . . . . . . . . . . . . . . . 21
    2.6 Simplification of the model for analysis . . . . . . . . . . . . . 22

**3 Trajectory tracking control**     25
    3.1 Introduction . . . . . . . . . . . . . . . . . . . . . . . . . . . . 25
    3.2 Analysis of the control system . . . . . . . . . . . . . . . . . . 26
    3.3 State of the art . . . . . . . . . . . . . . . . . . . . . . . . . . 29
    3.4 Introduction to nonlinear predictive control . . . . . . . . . . . 35
    3.5 Linear control of the inner loop . . . . . . . . . . . . . . . . . 37
    3.6 Nonlinear predictive control of the outer loop . . . . . . . . . . 39

**4 Navigation**     45
    4.1 Introduction . . . . . . . . . . . . . . . . . . . . . . . . . . . . 45
    4.2 State of the art . . . . . . . . . . . . . . . . . . . . . . . . . . 46
    4.3 Available sensors . . . . . . . . . . . . . . . . . . . . . . . . . 47
    4.4 Introduction to the developed algorithm . . . . . . . . . . . . 49
    4.5 State estimation . . . . . . . . . . . . . . . . . . . . . . . . . . 49
    4.6 Slip control . . . . . . . . . . . . . . . . . . . . . . . . . . . . 55
    4.7 Block diagram . . . . . . . . . . . . . . . . . . . . . . . . . . . 57

**5 Motion planning**     59
    5.1 Introduction . . . . . . . . . . . . . . . . . . . . . . . . . . . . 59
    5.2 State of the art . . . . . . . . . . . . . . . . . . . . . . . . . . 59
    5.3 Introduction to the developed algorithms . . . . . . . . . . . . 61
    5.4 Trajectory generation . . . . . . . . . . . . . . . . . . . . . . . 63
    5.5 Map-building . . . . . . . . . . . . . . . . . . . . . . . . . . . 72
    5.6 Waypoint generation . . . . . . . . . . . . . . . . . . . . . . . 73

| 6 | **Results** | **75** |
|---|---|---|
| | 6.1 Trajectory tracking | 76 |
| | 6.2 Point stabilisation | 77 |
| | 6.3 Navigation | 79 |
| | 6.4 Motion planning | 83 |

| 7 | **Conclusion and outlook** | **85** |
|---|---|---|

| A | **Technical specifications** | **87** |
|---|---|---|
| | A.1 Hardware | 87 |
| | A.2 Software | 91 |

| B | **The Extended Kalman Filter** | **93** |
|---|---|---|
| | B.1 Analytical derivation | 93 |
| | B.2 Practical issues | 95 |

**Bibliography**     **97**

# Chapter 1

# Introduction

## 1.1 Motivation

Research on autonomous mobile robots or vehicles is one of the most seminal subjects at the moment. Economically leading nations like the USA and Japan have adopted it to one of their national research topics. The reason for this considerable research interest is a number of possible applications in various fields. Autonomous transport devices could increase the degree of automation of complex production processes, thus significantly improving productivity. In case of natural disasters, autonomous vehicles could be deployed to dangerous environments without risking harm to human operators, and perform rescue missions. Furthermore, autonomous vehicles could be helpful in difficult construction problems as transport or operating devices in environments which are dangerous or difficult to access. Finally, an expanded degree of autonomy in automotive applications alleviates the burden on the driver and can increase safety. This is especially interesting for physically disabled drivers.

## 1.2 Scope

According to a definition found in Wikipedia [99], the term *autonomous* stems from the greek words *auto*, meaning *by oneself*, and *nomos*, meaning *law*. It can therefore be translated as *giving oneself his own law*, or behaviour free from external authority. This means that an autonomous robot must be capable of performing its mission without any interaction with a human operator. A still more restrictive interpretation, especially targeting the aspect of self-localisation, forbids the dependency on artificial external devices for fully autonomous operation.

Autonomous operation requires some implementation of artificial intelligence. This is, however, not the focus of this work. This work focuses on motion control for autonomous vehicles, comparable to the subconscious motoric capabilities of a human being, as opposed to the conscious and very abstract process of deciding over one's actions.
As the product of some superordinate decision-making process, a goal is specified. At this point, motion control comes into effect, which can be subdivided into motion planning, tracking control and navigation: A motion to reach the goal is planned (motion planning). This motion is executed (tracking control) while verifying the accurateness of its execution (navigation).

In the course of this work, for each of the specified subtasks algorithms were developed, using a tiny two-wheeled mobile robot called Tinyphoon. In the following those algorithms are presented and experimentally evaluated.

# Chapter 2

# Modelling

## 2.1 Introduction

A mathematical model of a dynamic system, in this context called plant, to be controlled is required for two reasons:
On the one hand it serves for a numerical simulation of the plant's behaviour, which allows the virtual assessment of developed controller concepts during the design process without risk to the hardware. Furthermore, the implementation of control algorithms in a simulation instead of on the hardware itself is usually much more convenient and less time-consuming. A simulation, however, cannot entirely replace thorough testing of a control concept on the hardware.
On the other hand, the model is the basis for mathematical analysis, which provides detailed information about the plant's dynamical properties and forms the starting point for conceptual controller design.
A mathematical model can never fully describe physical reality, no matter how complex it becomes. Therefore, its complexity and its level of accuracy is more or less subject to a trade-off between objective and expenditure.

Not many publications exclusively concerned with mobile robot modelling are found in the literature. Obviously most authors working in the field do not put much effort into modelling, usually wheel slippage is not included in their models. Among the publications explicitly addressing wheel slippage, which is the crucial difficulty in wheeled mobile robot modelling, are the works of Balakrishna and Ghosal [8] and Williams *et al.* [100], performing modelling of omnidirectional robots. Muir and Newman develop a formalism in [69] to model mobile robots with an arbitrary number of wheels.
To find more elaborate friction models, the literature on vehicle dynamics has to be consulted. Liu and Peng for example present the so-called brush model in [65]. The famous magic formula model is described by Pacejka in [76].

## 2.2 Mathematical description

The robot, which is depicted in Fig. 2.1, has two wheels with rubber tyres and two felt shoes, one at the front and one at the rear to stabilise it around the pitch axis. The two wheels are supported by ball bearings and powered by two individual DC-motors via one-stage transmissions. A microcontroller ($\mu$C) produces two pulse-width modulated (PWM) constant voltage signals, which are amplified by a dual full bridge driver and applied to the two DC-motors. The algorithms are executed on the digital signal processor (DSP) and the resulting PWM-duty cycle values are sent to the $\mu$C for execution. The $\mu$C also collects the sensor signals and provides them to the DSP.

Detailed technical specifications of the robot hardware can be found in Appendix A.

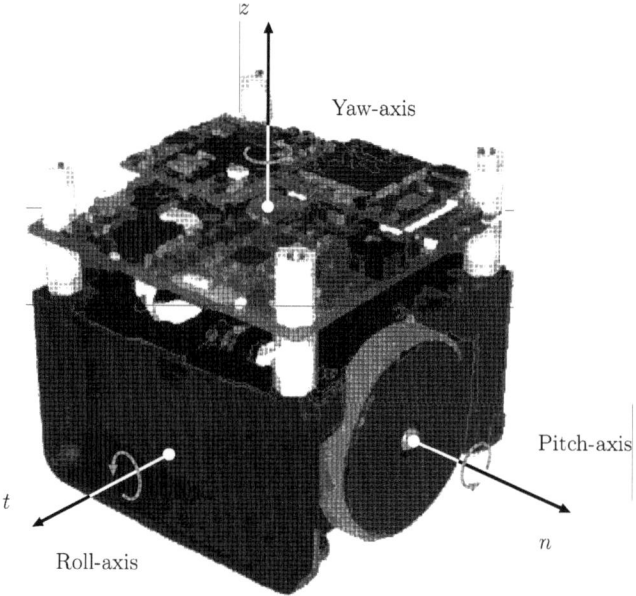

Figure 2.1: Mini-robot Tinyphoon.

### 2.2.1 Mechanical system component

The mechanical part of the robot consists of the wheels, the chassis, the ball bearings, the transmissions and the DC-motors. A local coordinate system is introduced as depicted in Fig. 2.1.

In the following, the symbols as listed in Table 2.1 are used.

A number of simplifying assumptions are made:

## 2.2. MATHEMATICAL DESCRIPTION

Table 2.1: Symbols

| Symbol | Description | Unit |
|---|---|---|
| $x, y$ | inertial coordinates | m |
| $\varphi$ | orientation | rad |
| $m$ | mass of the chassis plus both wheels | kg |
| $I$ | moment of inertia of the chassis | kgm$^2$ |
| $a$ | width and length of the robot's chassis | m |
| $b$ | distance between the two wheels' contact points | m |
| $s$ | position of the center of gravity in $z$-direction | m |
| $g$ | gravitational constant | ms$^{-2}$ |
| $F$ | force | N |
| $M$ | moment | Nm |
| $v$ | velocity | ms$^{-1}$ |
| $\omega$ | angular velocity | rads$^{-1}$ |
| $r$ | radius of a wheel | m |
| $m_\mathrm{w}$ | mass of a wheel | m |
| $I_\mathrm{w}$ | moment of inertia of a wheel | kgm$^2$ |
| $I_\mathrm{m}$ | moment of inertia of a DC-motor | kgm$^2$ |
| $I_\mathrm{w,tot.}$ | moment of inertia of a wheel plus the reduced moment of inertia of a DC-motor | kgm$^2$ |
| $n$ | gear ratio | 1 |
| $B, C, D, E$ | parameters for a simplified magic formula model [76]. | various |
| ~~$s_\mathrm{t}$~~ | ~~tangential wheel-slip~~ | ~~1~~ |
| $K_\mathrm{t}, K_\mathrm{n}$ | friction coefficients of the felt-shoes in tangential and normal direction | Nsm$^{-1}$ |
| R | index for *right* | |
| L | index for *left* | |
| t | index for *tangential* | |
| n | index for *normal* | |
| N | index for *normal to the surface* | |
| w | index for *wheel* | |
| m | index for *motor* | |
| f | index for *friction* | |

1. The chassis is treated as a homogeneous cuboid, its moment of inertia is calculated from the edge lengths and its mass. The position of the center of gravity, however, is measured.

2. The wheels, too, are treated as homogeneous cylinders, their moment of inertia is calculated as above.

3. It is assumed that there is no rotation around the pitch-axis, i.e. the felt shoes are free of play.

4. The wheels stay in contact with the surface at all times, i.e. the contact forces are always greater than zero. The weight of the robot is entirely carried by the wheels' contact

surfaces, the contact forces of the felt shoes are assumed to be zero. Nevertheless, it is assumed that the felt shoes produce linearly velocity-dependent friction forces.

5. There is no coupling between the tangential slip and the normal slip of the wheels. The tangential force depends on the tangential slip $s_{t,R}$ and $s_{t,L}$, respectively, whereas the normal force depends on the side-slip velocity $v_n$.

6. The normal force $F_n$ can be calculated without respect to the individual contact forces for the cumulative contact area of both wheels.

The relevant dynamic and kinematic quantities are introduced in Fig. 2.2.

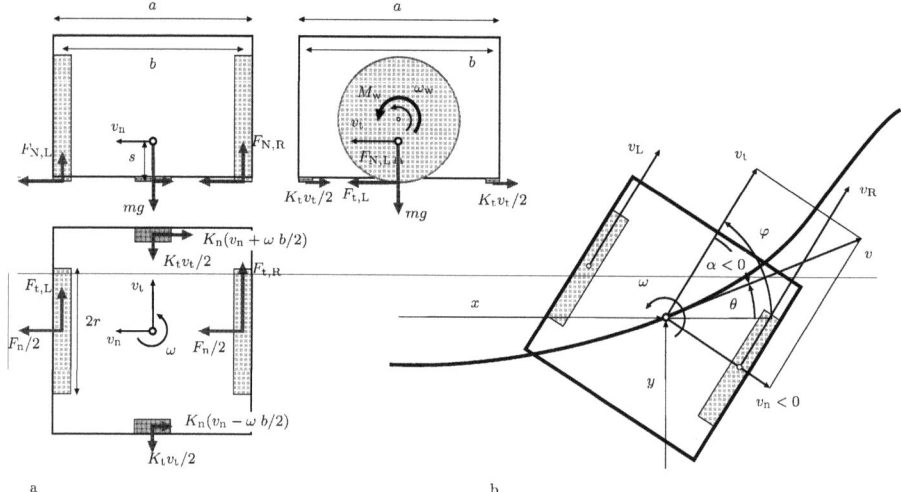

Figure 2.2: Kinematic and dynamic quantities and geometric dimensions of the robot.

The kinematics of the robot are given by

$$\dot{x} = v_t \cos\varphi - v_n \sin\varphi$$
$$\dot{y} = v_t \sin\varphi + v_n \cos\varphi \qquad (2.1)$$
$$\dot{\varphi} = \omega$$

which is equivalent to

$$\dot{x} = v\cos(\varphi + \alpha) = v\cos\theta$$
$$\dot{y} = v\sin(\varphi + \alpha) = v\sin\theta \qquad (2.2)$$
$$\dot{\varphi} = \omega$$

via $v_t = v\cos\alpha$ and $v_n = v\sin\alpha$ and the theorems of addition of the trigonometric functions.

## 2.2. MATHEMATICAL DESCRIPTION

To describe the motion of the system, the principles of conservation of translational and angular momentum are written in the local body-fixed frame of the robot. Therefore, the absolute accelerations are needed.
Differentiating (2.1) yields

$$\begin{aligned}\ddot{x} &= \dot{v}_t \cos\varphi - v_t \sin\varphi\,\omega - \dot{v}_n \sin\varphi - v_n \cos\varphi\,\omega \\ \ddot{y} &= \dot{v}_t \sin\varphi + v_t \cos\varphi\,\omega + \dot{v}_n \cos\varphi - v_n \sin\varphi\,\omega,\end{aligned} \quad (2.3)$$

which can be rearranged into

$$\begin{bmatrix}\ddot{x}\\\ddot{y}\end{bmatrix} = \underbrace{\begin{bmatrix}\cos\varphi & -\sin\varphi\\\sin\varphi & \cos\varphi\end{bmatrix}}_{T}\begin{bmatrix}\dot{v}_t - v_n\omega\\\dot{v}_n + v_t\omega\end{bmatrix}. \quad (2.4)$$

Since the matrix $T$ performs the rotational transformation from the local into the global frame, the absolute accelerations in the local frame are found to be

$$\begin{bmatrix}a_t\\a_n\end{bmatrix} = \begin{bmatrix}\dot{v}_t - v_n\omega\\\dot{v}_n + v_t\omega\end{bmatrix}. \quad (2.5)$$

Conservation of translational momentum in t-direction then yields

$$\dot{v}_t = \frac{1}{m}\left(F_{t,R} + F_{t,L} - K_t v_t\right) + v_n\omega, \quad (2.6)$$

and in n-direction

$$\dot{v}_n = \frac{1}{m}\left(F_n - 2K_n v_n\right) - v_t\omega. \quad (2.7)$$

Conservation of angular momentum around the z-axis is written as

$$\dot{\omega} = \frac{b}{2I}\left(F_{t,R} - F_{t,L} - K_n b\,\omega\right), \quad (2.8)$$

where the polar moment of inertia is found as $I = ma^2/6$.
With these equations, the three degrees of freedom of planar motion of a rigid body are covered. Under the given assumptions, the robot exhibits two more degrees of freedom, that is to say the rotation of both wheels. The principle of conservation of angular momentum for each of the two wheels reads

$$\dot{\omega}_w = \frac{1}{I_{w,tot.}}(M_m - M_f - F_t r), \quad (2.9)$$

where $I_{w,tot.} = I_w + I_m n^2 = m_w r^2/2 + I_m n^2$.
To complete the equations of motion, expressions for the forces $F_n$, $F_{t,R}$ and $F_{t,L}$ and the friction moment $M_f$ acting on the system have to be found. At this stage, the motor moments are regarded as the inputs to the system. The normal force $F_n$ acting on the wheels' consummate contact area is assumed to be a function of the side-slip velocity and modelled using the basic structure of the well-known magic formula [76],

$$F_n = -mgD\sin(C\operatorname{atan}(Bv_n - E(Bv_n - \operatorname{atan}(Bv_n)))). \quad (2.10)$$

The tangential forces $F_{t,R}$ and $F_{t,L}$ are assumed to depend on the contact force of the respective wheel and the wheel slip. The contact forces $F_{N,R}$ and $F_{N,L}$ are found by combining equilibrium of moments around the roll-axis and equilibrium of forces in $z$-direction,

$$F_{N,R} = \frac{mg}{2} - \frac{s}{b}(F_n + 2K_n v_n)$$
$$F_{N,L} = \frac{mg}{2} + \frac{s}{b}(F_n + 2K_n v_n)$$
(2.11)

The tangential wheel slip is defined as the relative difference between the wheel's circumferential velocity and the velocity of its hub,

$$s_{t,R} = \frac{\omega_{w,R} r - (v_t + \omega b/2)}{|v_t + \omega b/2|}$$
$$s_{t,L} = \frac{\omega_{w,L} r - (v_t - \omega b/2)}{|v_t - \omega b/2|}.$$
(2.12)

To avoid problems in the numerical simulation due to the singularity at $|v_t \pm \omega b/2| \to 0$, (2.12) is replaced by

$$s_{t,R} = \frac{\omega_{w,R} r - (v_t + \omega b/2)}{\delta}$$
$$s_{t,L} = \frac{\omega_{w,L} r - (v_t - \omega b/2)}{\delta}.$$
(2.13)

whenever $|v_t \pm \omega b/2| < \delta$. Moreover, the absolute value of the slip is limited to one.
The tangential forces as a function of tangential slip $s_t$ and the contact forces are formulated using the magic formula,

$$F_t = \max(F_N, 0) D \sin(C \operatorname{atan}(Bs_t - E(Bs_t - \operatorname{atan}(Bs_t)))),$$
(2.14)

where the parameters $B$ through $E$ are different from those used in (2.10).
The magic formula characteristics are depicted in Fig. 2.3.

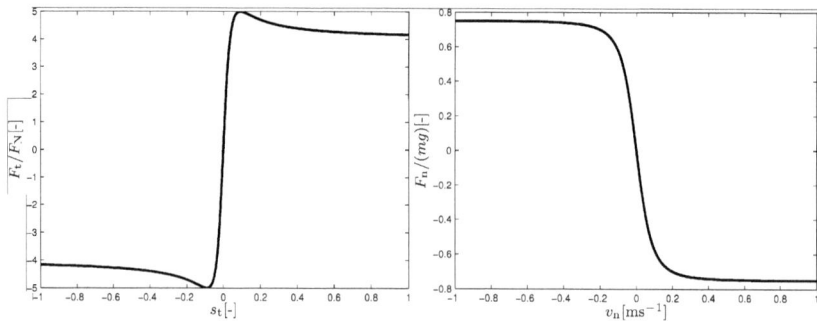

Figure 2.3: Magic formula model for tangential and lateral slip.

## 2.2. MATHEMATICAL DESCRIPTION

To obtain a sufficiently accurate but simple model for the friction moment $M_f$ a characteristic containing linear and parabolic sections is fitted to a number of data points obtained by stationary measurement, as illustrated in Fig. 2.4.

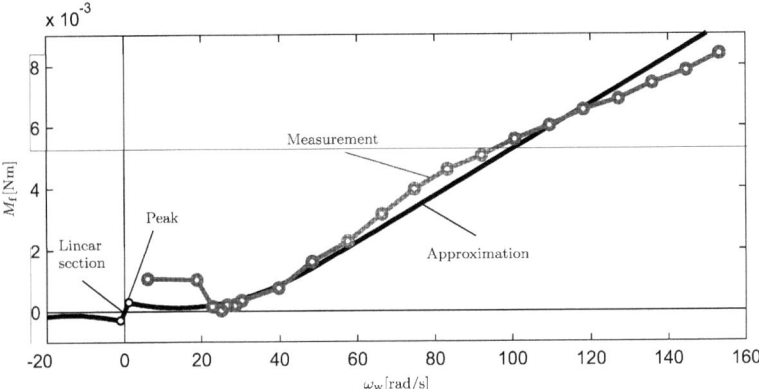

Figure 2.4: Bearing friction moment characteristic.

To avoid problems in the numerical simulation, the discontinuity at $\omega_w = 0$ is replaced by a linear section, and the peaks are reduced.

Inserting equations (2.10) through (2.14) into the differential equations (2.6) through (2.9) yields a closed system of five differential equations for five degrees of freedom.
Finally, the inertial coordinates and the orientation are calculated by integrating (2.1).

### 2.2.2 Electrical system component

The robot's DC-motors are driven by pulse width modulated (PWM) voltage signals with constant amplitude. To characterise a PWM-signal the proportion between the pulse width and the periodic time $r_{PWM} = T_{on}/T_{PWM}$ is used. Since the original signals are generated by the microcontroller, which cannot supply sufficient energy to actually drive the DC-motors, the signals are amplified by a dual full bridge driver.
The symbols used in this section and their interpretations are listed in Table 2.2.
Fig. 2.5 shows the electrical circuit to be modelled. Technical specifications are to be found in Appendix A.
For the electrical model, the following simplifying assumptions are made:

1. The internal resistance of the voltage source is neglected, i.e. the supply voltage is assumed constant.

2. The voltage drops of the transistors and the diodes are assumed constant.

3. The drift of the system due to temperature changes is neglected.

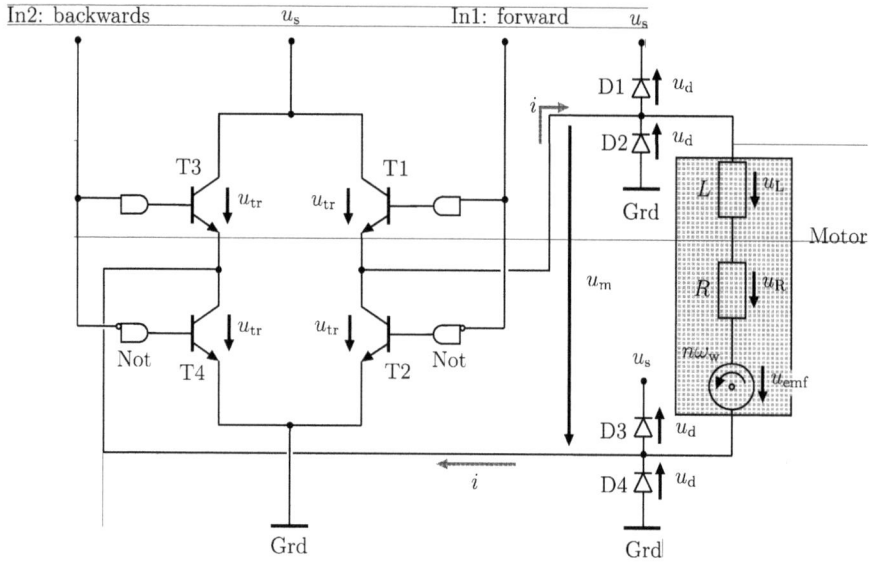

Figure 2.5: Electrical circuit with transistors T1-T4 and diodes D1-D4.

Table 2.2: Symbols

| Symbol | Description | Unit |
|---|---|---|
| $u$ | voltage | V |
| $i$ | current | A |
| $n$ | transmission ratio | 1 |
| $R$ | resistance | $\Omega$ |
| $L$ | inductance | $\Omega$s |
| ~~$K$~~ | ~~electromagnetic force (emf) constant or motor constant~~ | ~~Vs or NmA$^{-1}$~~ |
| $T_{\text{PWM}}$ | PWM-period | s |
| $T_{\text{on}}$ | active part of the PWM-period | s |
| $r_{\text{PWM}}$ | PWM-ratio | 1 |
| tr | index for *transistor* | |
| d | index for *diode* | |
| emf | index for *electromagnetic force* | |
| s | index for *supply* | |
| R | index for *resistance* | |
| L | index for *inductance* | |
| m | index for *motor* | |

## 2.2. MATHEMATICAL DESCRIPTION

4. Capacities and inductances of the elements in the circuit are neglected (except for the inductance of the motor-coil).

5. The switch-on and switch-off time of the transistors are assumed constant but different. The diodes are modelled to react instantaneously.

The dynamics of each DC-motor are described by Kirchhoff's law, introducing the current $i$ as additional degree of freedom for each wheel,

$$\frac{di}{dt} = \frac{1}{L}(u_\mathrm{m} - Ri - Kn\omega_\mathrm{w}). \tag{2.15}$$

The relationship between the current and the moment produced by the motor is linear,

$$M_\mathrm{m} = Kni. \tag{2.16}$$

Due to the structure of the circuit, the terminal voltage of the motor $u_\mathrm{m}$ assumes several different magnitudes as a result of the changing direction of the current and the changing transistor states. Basically, there are four different scenarios, here illustrated for a PWM-signal at In1 and low at In2, this means that the motor is driven in forward direction. For the opposite direction the labels of the diodes and transistors would have to be exchanged with their symmetric counterparts (T1 with T3, T2 with T4) in the following description.

1. T1 and T4 are conductive, T2 and T3 are non-conductive and $i > 0$. This is normally the case at the end of the active part of the PWM-period, when the wheel is actually driven by the motor. Then the current passes through T1, the motor and T3. The terminal voltage in this case is $u_\mathrm{m} = u_\mathrm{s} - 2u_\mathrm{tr}$.

2. T2 and T4 are conductive, T1 and T3 are non-conductive and $i > 0$. This is normally the case at the beginning of the passive part of the PWM-period. The motor works as a generator in this case. Here, the current passes through D2, the motor and T3. The terminal voltage then results to $u_\mathrm{m} = -u_\mathrm{d} - u_\mathrm{tr}$.

3. T2 and T4 are conductive, T1 and T3 are non-conductive and $i > 0$. This situation occurs during the passive part of the PWM-period, after the generator-effect of the motor turns the current over to the opposite direction. Then the current goes through D4, the motor and T2. The terminal voltage becomes $u_\mathrm{m} = u_\mathrm{d} + u_\mathrm{tr}$.

4. T1 and T4 are conductive, T2 and T3 non-conductive, $i > 0$. This happens at the beginning of the active part of the PWM-period, when the current is still negative. The current then passes through D4, the motor and D1. The terminal voltage becomes $u_\mathrm{m} = u_\mathrm{s} + 2u_\mathrm{d}$.

However, there are some more effects which lead to a number of additional scenarios:

1b. Since the switch-on time of the transistors is larger than the switch-off time the following effect can occur: At the beginning of the active part of the PWM-period T2 becomes non-conductive, but T1 is not yet conductive. If in this situation the current becomes already positive, the circuit is interrupted, both $u_\mathrm{m}$ and $i$ are equal to zero.

This ends as soon as T1 becomes conductive, then scenario 1 comes into effect.

All the situations so far are illustrated in the left part of Fig. 2.6.

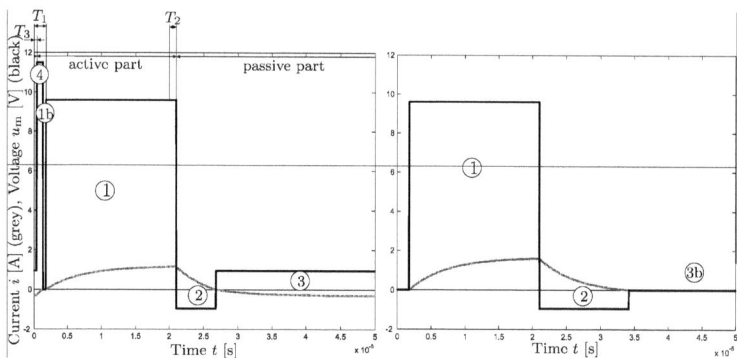

Figure 2.6: Current and Voltage during a PWM-period, 1-4 and 1b, left, and 1-3b, right. $T_1$ denotes the switch-on time of a transistor, $T_2$ the switch-off time after high voltage and $T_3$ the switch-off time after low voltage.

3b. For scenario 3 a different situation can arise: If $u_m > u_{cmf}$, the current grows, but then scenario 2 comes into effect and the current decreases, asf. The result is a high-frequency low-amplitude oscillation, which is modelled by holding the current equal to zero as shown in the right part of Fig. 2.6.

1.,2. If the angular velocity is smaller than a certain margin (or negative) the current will not cross zero, it will remain positive during the whole PWM-period. In this case, only scenarios 1 and 2 come into effect, as illustrated in the right part of Fig. 2.7.

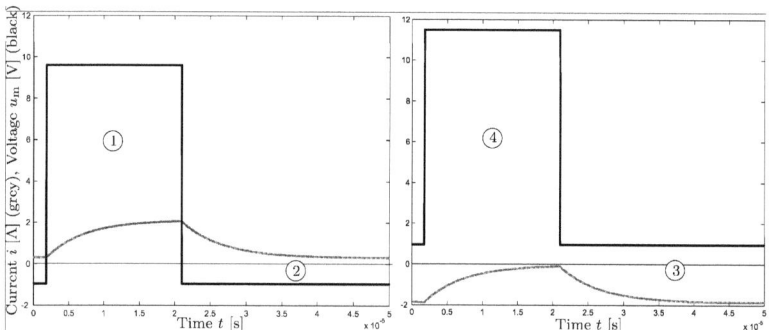

Figure 2.7: Current and Voltage during a PWM-period, 1 and 2, left, and 3 and 4, right.

3.,4. If the angular velocity is greater than a certain margin, the current will not cross zero, too. It will remain negative throughout the period such that only scenarios 3 and 4 take place, see the left part of Fig. 2.7.

## 2.3 Numerical simulation

By combining the electrical and the mechanical model, the movement of the robot as a response to two PWM-input signals can be simulated. The combined dynamics, which comprise seven degrees of freedom, exhibit time constants of strongly differing orders of magnitude: electrodynamics $O(10^{-5}\text{s})$, mechanics $O(10^0\text{s})$. Therefore, a complete numerical simulation is very inefficient. However, the two parts can be decoupled by the following assumptions:

- The wheels' angular velocities are assumed to remain constant during a PWM-period, which is justified because the wheels' dynamics are much slower than the electrodynamics.

- The current at the end of a PWM-period is assumed to be equal to the current at the beginning, i.e. the system is assumed to be stationary. This assumption is very accurately fulfilled as long as the PWM-input is continuous. If it is not continuous, the stationarity assumption is violated at most for a few PWM-periods after the discontinuity.

Then the integral mean value of the current, and thereby the moment over one PWM-period can be calculated analytically according to the different scenarios. The integral mean value of the moment is then a function of the PWM-ratio $r_{\text{PWM}}$, the wheel angular rate $\omega_{\text{w}}$ and the supply voltage $u_{\text{s}}$ as a parameter.

The numerical simulation is thereby reduced to the five mechanical degrees of freedom with the two motor moments as inputs. By this method the calculation time is decreased by a factor of around 100, while no difference to the full simulation can be observed.

## 2.4 Measurements

A number of physical parameters have to be adjusted. Some can be measured directly, see Appendix A, others have to be obtained indirectly, i.e. they have to be calculated from other results or in the worst case be found by systematic variation and validation. In Table 2.3 the physical parameters and their respective measurement methods are listed.

Table 2.3: Measurement methods

| Quantity | Method | Direct/indirect |
|---|---|---|
| mass | weighing | direct |
| radius/edge length | measuring | direct |
| moment of inertia | weighing/measuring | indirect: calculated from mass and lengths |
| position of the center of gravity | balancing | direct |
| transmission ratio | known from design | direct |
| inductance | oscilloscope and signal generator | direct |
| resistance | multimeter | direct |
| switching times | oscilloscope | direct |
| diode and transistor voltage drops | oscilloscope | direct |
| supply voltage | multimeter | direct |
| bearing friction moment | measurement of unloaded stationary wheel angular velocities | indirect: calculated from integral mean of the current |
| side-slip and tangential slip parameters | video processing system | indirect: adjustment and validation |
| dynamic friction parameters | video processing system | indirect: adjustment and validation |

### 2.4.1 Video measurement

To obtain data describing the movement of the robot for validation purposes, a video measurement system is developed. The CCD-camera records a video of the moving robot from above at 25 frames per second (fps), corresponding to a sampling time of $T_V = 0.04$s, in 640×480pixels .avi-format. Detailed information about hardware and software used for the video measurement system is compiled in Appendix A.

The video is decomposed into its individual frames in 24bit bitmap (8bit each for red, green and blue colour channel) format and subsequently processed.

To measure position and orientation of the robot in 2D, at least two points on top of the robot must be identified in each frame. Therefore, the robot is covered by a colour marker with two rectangular panels of different colour (red and white), which can be identified by differential keying: A reference image is subtracted from the current image such that the two rectangles

## 2.4. MEASUREMENTS

are the remaining features, see Fig. 2.8. The white marker appears equally on all three colour channels, the red marker predominantly on the red channel.

Figure 2.8: Original image (a), red channel with centroid and correction (b), blue and green channel with centroid of the white marker and correction (c) and original image overlayed with measurements (d). The lower left corner of the image corresponds to the camera's central axis.

Then all colour values exceeding an upper bound are equalised and all colour values below a lower bound are replaced by zero. The centroid of the white marker is calculated from the arithmetic mean of the green and the blue channel, the centroid of the overall area is calculated from the red channel.

The colour strength of each of the three colour channels, an 8bit integer for each pixel is arranged in three matrices, $R, G, B \in \mathbb{R}^{480 \times 640}$ in a left-handed coordinate system, i.e. the $y$-coordinate is mirrored and the origin of the coordinate system is in the upper left corner of the image. Then the left-handed coordinates of the centroid of a colour channel, exemplarily written for the red channel, are given by

$$x_\text{R} = \Big( \underbrace{1^\text{T}}_{[1\times 480]} \underbrace{R}_{[480\times 640]} \underbrace{[1\ldots 640]^\text{T}}_{[640\times 1]} \Big) \Big/ \Big( \underbrace{1^\text{T}}_{[1\times 480]} \underbrace{R}_{[480\times 640]} \underbrace{1}_{[640\times 1]} \Big)$$
$$y_\text{R} = \Big( \underbrace{1^\text{T}}_{[1\times 640]} \underbrace{R^\text{T}}_{[640\times 480]} \underbrace{[1\ldots 480]^\text{T}}_{[480\times 1]} \Big) \Big/ \Big( \underbrace{1^\text{T}}_{[1\times 480]} \underbrace{R}_{[480\times 640]} \underbrace{1}_{[640\times 1]} \Big). \quad (2.17)$$

Due to the deviation from the camera's central axis and the mounting height of the colour marker, the centroid coordinates are subject to a systematic error as shown in Fig. 2.9, which is corrected by applying the theorem on intersecting lines twice,

$$\begin{aligned} x &= \frac{H-h}{H}(\tilde{x} - 320) + 320 \\ y &= \frac{H-h}{H}(\tilde{y} - 240) + 240. \end{aligned} \quad (2.18)$$

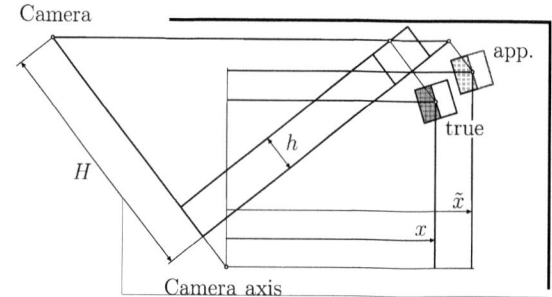

Figure 2.9: Systematic error due to the panel's mounting height $h$: Apparent and true location of the panel. $H$ denotes the distance between ground and camera.

Next, a scaling factor in meters per pixel, which is found by measuring the display window's dimensions, is used to obtain the coordinates of the robot's centroid $x$ and $y$ and the coordinates of the centroid of the white marker $x_\text{w}$ and $y_\text{w}$ in SI-units.
The orientation is incrementally calculated using the four-quadrant inverse tangent,

$$\varphi_1 = \text{atan2}(-y_\text{w1} + y_1, x_\text{w1} - x_1), \quad (2.19)$$

and

$$\varphi_i = \varphi_{i-1} + \text{atan2}(\widetilde{\Delta y_i}, \widetilde{\Delta x_i}), \quad (2.20)$$

where

$$\begin{bmatrix} \widetilde{\Delta x_i} \\ \widetilde{\Delta y_i} \end{bmatrix} = \begin{bmatrix} \cos\varphi_{i-1} & \sin\varphi_{i-1} \\ -\sin\varphi_{i-1} & \cos\varphi_{i-1} \end{bmatrix} \begin{bmatrix} x_{\text{w},i} - x_i \\ -y_{\text{w},i} + y_i \end{bmatrix}. \quad (2.21)$$

The integer $i$ denotes the currently processed frame and the index 1 denotes the first frame. The sign of $y$ is inverted to compensate the original left-handed coordinate system. For each

## 2.4. MEASUREMENTS

frame, the vector pointing from the robot's centroid towards the centroid of the white marker is transformed into the local coordinate system of the previous frame, the orientation angle increment is calculated and added to the orientation angle of the previous frame. Thereby the orientation angle evolves on R, instead of being limited to the definition interval of the four-quadrant inverse tangent, $[-\pi, \pi]$, and thereby possibly exhibiting discontinuities. This procedure works under the reasonable assumption that no rotation through more than $\pi$ radians takes place between two subsequent frames.

Finally, position and orientation are transformed such that the initial posture of the robot represents the origin of the coordinate system,

$$\begin{bmatrix} x_i \\ y_i \\ \varphi_i \end{bmatrix} = \begin{bmatrix} \cos \varphi_{C,1} & \sin \varphi_{C,1} & 0 \\ -\sin \varphi_{C,1} & \cos \varphi_{C,1} & 0 \\ 0 & 0 & 1 \end{bmatrix} \begin{bmatrix} x_{C,i} - x_{C,1} \\ -y_{C,i} + y_{C,1} \\ \varphi_{C,i} - \varphi_{C,1} \end{bmatrix}, \qquad (2.22)$$

where the index C denotes the original frame of reference of the camera. Again, the sign of $y$ is inverted.

To obtain the velocity components $v_\text{n}$ and $v_\text{t}$ and the yaw rate $\omega$, the coordinates $x$ and $y$ and the orientation $\varphi$ are approximated piecewise by quadratic splines, which are subsequently differentiated analytically. In the sequel the procedure is illustrated for the coordinate $x$.

An approximation by quadratic spline is given by

$$x = p_0 + p_1 \tau + p_2 \tau^2. \qquad (2.23)$$

To obtain the derivative at position $i$, a spline is calculated using values at positions $i-1$, $i$ and $i+1$, $\tau$ is defined on the interval $[-1; 1]$. Then the parameters are calculated according to

$$\begin{bmatrix} p_0 \\ p_1 \\ p_2 \end{bmatrix} = \begin{bmatrix} 1 & -1 & 1 \\ 1 & 0 & 0 \\ 1 & 1 & 1 \end{bmatrix}^{-1} \begin{bmatrix} x_{i-1} \\ x_i \\ x_{i+1} \end{bmatrix}, \qquad (2.24)$$

and the derivative is obtained as

$$\dot{x}_i = \dot{x}(\tau = 0) = p_1 \frac{1}{T_\text{V}}. \qquad (2.25)$$

The derivatives at the initial position $i = 1$ are assumed to be zero, i.e. the robot is assumed to be initially at rest, whereas the derivatives at the final position are set equal to those at the second to last position.

The path angle $\theta$ is calculated using the same incremental procedure as for the orientation angle in (2.19) through (2.21),

$$\theta = \operatorname{atan2}(\dot{y}, \dot{x}). \qquad (2.26)$$

The velocity components are given by transformation into the local frame

$$\begin{aligned} v_{\text{t},i} &= \cos(\theta_i - \varphi_i)\sqrt{\dot{x}^2 + \dot{y}^2} \\ v_{\text{n},i} &= \sin(\theta_i - \varphi_i)\sqrt{\dot{x}^2 + \dot{y}^2} \\ \omega_i &= \dot{\varphi}_i. \end{aligned} \qquad (2.27)$$

The difference $\theta_i - \varphi_i$ is defined on $\mathbb{R}$, its magnitude also depends on the direction of motion, whereas the side-slip angle is only defined on the interval $[-\frac{\pi}{2}, \frac{\pi}{2}]$. The side-slip angle is calculated according to

$$\alpha_i = \operatorname{atan}\left(\frac{v_{\mathrm{n},i}}{v_{\mathrm{t},i}}\right) f\left(\sqrt{\dot{x}_i^2 + \dot{y}_i^2}, \omega_i\right) \tag{2.28}$$

with the linear validity function

$$f\left(\sqrt{\dot{x}_i^2 + \dot{y}_i^2}, \omega_i\right) = \begin{cases} 0 & : \quad \sqrt{\dot{x}^2 + \dot{y}^2}|\omega_i| < c \\ \dfrac{\sqrt{\dot{x}^2 + \dot{y}^2}|\omega_i| - c}{d - c} & : \quad c < \sqrt{\dot{x}^2 + \dot{y}^2}|\omega_i| < d \\ 1 & : \quad d < \sqrt{\dot{x}^2 + \dot{y}^2}|\omega_i| \end{cases} \tag{2.29}$$

enforcing the side-slip angle to vanish for low track speed and/or low curvature, where otherwise no meaningful results can be obtained due to numerical reasons, and the side-slip can reasonably be assumed to be very small.

The parameters $c$ and $d$ are found by systematic adjustment.

## 2.5 Validation

After adjustment of the parameters, the model is validated by applying a random sequence of PWM-inputs to each motor. The resulting movement is recorded and compared with the simulation results for the same inputs, see Fig. 2.10 and Fig. 2.11. The wheel angular velocities as measured by the wheel encoders are stored in a log-file on the microcontroller and are also used for validation.

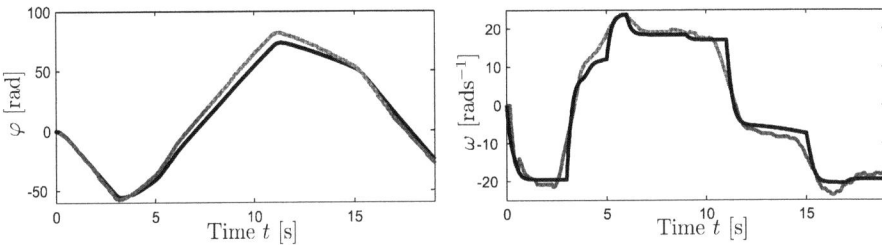

Figure 2.10: Model validation: Orientation angle (left) and yaw rate (right), simulation in black and video measurement in grey.

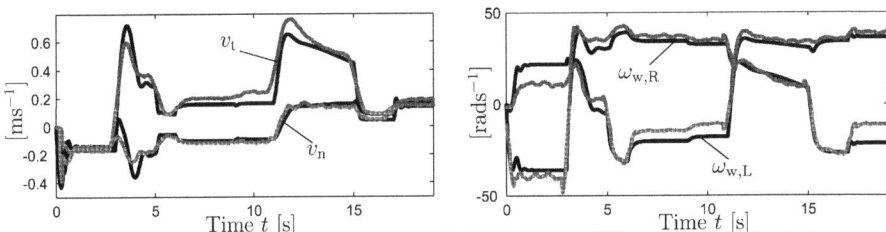

Figure 2.11: Model validation: Tangential and normal velocity components (left) and wheel angular velocities (right), simulation in black and measurement in grey.

Individual validation of the model of the electrical circuit is carried out by measuring the PWM input signal, the motor voltage $u_\mathrm{m}$ and the current $i$ via a serial resistor in steady state using a standard four-channel oscilloscope, see Fig. 2.12.

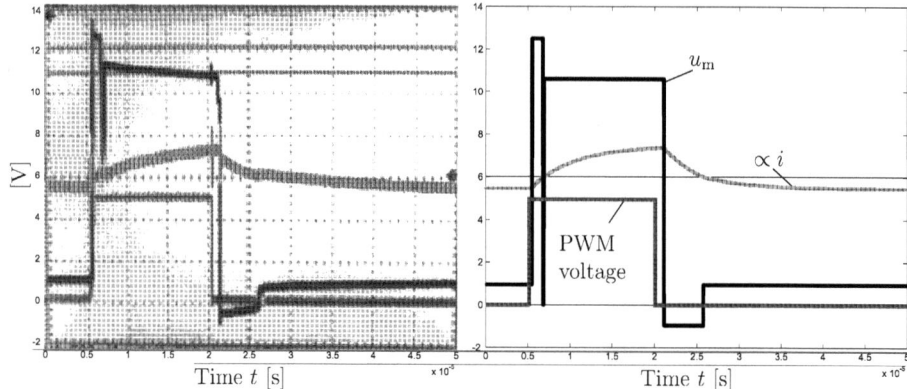

Figure 2.12: Model validation: Input voltage, motor voltage and current-proportional voltage with zero at 6V for a PWM-ratio of 0.3. Left: Screenshot of oscilloscope. Right: Simulation.

## 2.6 Simplification of the model for analysis

To obtain a closed-form analytical model description which is suitable for controller design, the original model derived for numerical simulation is further simplified by undertaking the following steps:

1. The rolling condition in tangential direction is assumed to be fulfilled, i.e. the tangential slip is neglected.

2. The dynamics of the side-slip velocity (2.7) are neglected.

3. In (2.6) the tangential velocity $v_t$ is replaced by the resulting linear velocity $v$, thereby assuming that the side-slip angle remains sufficiently small.

4. To have a model which is as generic as possible, the friction parameters $K_n$ and $K_t$ concerning individual ground conditions are discarded.

5. The relationship between the integral mean of the motor moment reduced by the friction moment $M_w = M_m - M_f$ and the PWM-ratio and the wheel angular velocity is numerically linearised to form a static bilinear characteristic.

6. The rate of change of the side-slip angle is assumed to be small, therefore $\dot{\theta} \approx \dot{\varphi} = \omega$ instead of $\dot{\theta} = \dot{\varphi} + \dot{\alpha}$.

With assumption 1 the five remaining degrees of freedom are reduced to three, for both wheels (2.9) is replaced by

$$F_t = \frac{M_m - M_f}{r} = \frac{M_w}{r} \qquad (2.30)$$

## 2.6. SIMPLIFICATION OF THE MODEL FOR ANALYSIS

and

$$\omega_{w,R} = \frac{v + b\,\omega/2}{r}$$
$$\omega_{w,L} = \frac{v - b\,\omega/2}{r}.$$
(2.31)

Assumption 2 eliminates one more degree of freedom. According to step 3 and 4 the dynamics of the resulting track speed $v = \sqrt{v_t^2 + v_n^2}\,\text{sign}v_t$ as the first remaining degree of freedom are described by

$$\dot{v} = \frac{1}{m_{\text{tot.}}}(F_{t,R} + F_{t,L}),$$
(2.32)

where $m_{\text{tot.}} = m + \frac{2}{r^2}I_{w,\text{tot.}}$ denotes the mass of the entire vehicle extended by the equivalent mass of the polar moment of inertia of wheels and motor, $I_{w,\text{tot.}} = I_w + I_m n^2$. By the definition of the resulting track speed using the sign of the tangential velocity it is implicitly assumed that the side-slip angle $\alpha$ is bounded in the interval $[-\frac{\pi}{2}; \frac{\pi}{2}]$.

With step 4 the dynamics of the yaw rate $\omega$ as the second degree of freedom are written as

$$\dot{\omega} = \frac{b}{2I_{\text{tot.}}}(F_{t,R} - F_{t,L}),$$
(2.33)

with $I_{\text{tot.}} = I + \frac{b^2}{2r^2}I_{w,\text{tot.}}$, the moment of inertia of the vehicle extended by the wheels' and motors' contribution.

Step 5 is performed by application of the well-known least-squares approximation. The desired structure of the approximate linear relationship is

$$M_w = M_w(r_{\text{PWM}}, \omega_w) = p_1 r_{\text{PWM}} + p_2 \omega_w,$$
(2.34)

with the two unknown parameters $p_1$ and $p_2$. A constant term $p_0$ can be omitted, since from the more complex original model it is already known that the characteristic is skew-symmetric. The parameter estimation problem is formulated as

$$\underbrace{\begin{bmatrix} r_{\text{pwm},1} & \omega_1 \\ \vdots & \vdots \\ r_{\text{pwm},N} & \omega_N \end{bmatrix}}_{X} \underbrace{\begin{bmatrix} p_1 \\ p_2 \end{bmatrix}}_{p} = \underbrace{\begin{bmatrix} M_{w,1} \\ \vdots \\ M_{w,N} \end{bmatrix}}_{y} + e,$$
(2.35)

where $X$ denotes the regressor matrix, $y$ the output vector, $p$ the parameter vector and $e$ the residuals. Then the parameters $p$ that yield minimum sum-of-squared-residuals $e^T e$ are given by

$$p = (X^T X)^{-1} X^T y.$$
(2.36)

In Fig. 2.13 a comparison between the original model and the bilinear characteristic is depicted. Combining (2.30) through (2.34) yields

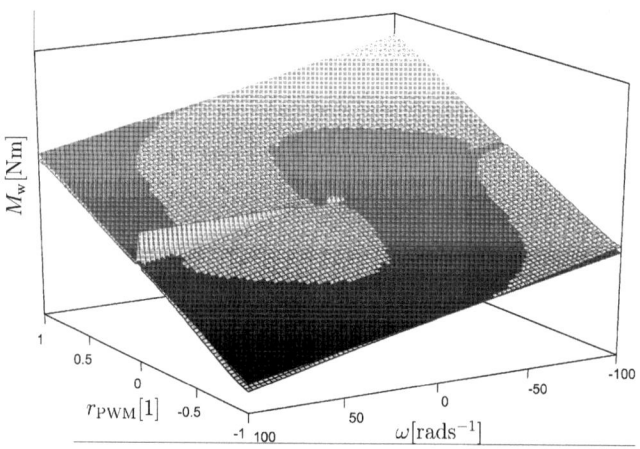

Figure 2.13: Original model (mesh) and bilinear characteristic (solid).

$$\begin{aligned}\dot{v} &= \frac{2p_2}{r^2 m_{\text{tot.}}} v + \frac{p_1}{r m_{\text{tot.}}} r_{\text{PWM,R}} + \frac{p_1}{r m_{\text{tot.}}} r_{\text{PWM,L}} \\ \dot{\omega} &= \frac{p_2 b^2}{2r^2 I_{\text{tot.}}} \omega + \frac{p_1 b}{2r I_{\text{tot.}}} r_{\text{PWM,R}} - \frac{p_1 b}{2r I_{\text{tot.}}} r_{\text{PWM,L}},\end{aligned} \quad (2.37)$$

rewritten as a continuous time state space system with state vector $\boldsymbol{v} := [v \ \omega]^{\text{T}}$, output vector $\boldsymbol{v}$ and input vector $\boldsymbol{u} := [r_{\text{PWM,R}} \ r_{\text{PWM,L}}]^{\text{T}}$

$$\dot{\boldsymbol{v}} = \underbrace{\begin{bmatrix} \frac{2p_2}{r^2 m_{\text{tot.}}} & 0 \\ 0 & \frac{p_2 b^2}{2r^2 I_{\text{tot.}}} \end{bmatrix}}_{\boldsymbol{A}} \boldsymbol{v} + \underbrace{\begin{bmatrix} \frac{p_1}{r m_{\text{tot.}}} & \frac{p_1}{r m_{\text{tot.}}} \\ \frac{p_1 b}{2r I_{\text{tot.}}} & -\frac{p_1 b}{2r I_{\text{tot.}}} \end{bmatrix}}_{\boldsymbol{B}} \boldsymbol{u}, \quad (2.38)$$

where $\boldsymbol{A}$ denotes the system matrix and $\boldsymbol{B}$ the input matrix. This system will be referred to as the robot's velocity dynamics. With step 6 (2.2) is approximately written as

$$\begin{aligned}\dot{x} &= v \cos \theta \\ \dot{y} &= v \sin \theta \\ \dot{\theta} &= \omega.\end{aligned} \quad (2.39)$$

By this system of equations, in the following referred to as the robot's kinematics, the motion of an imaginary vehicle without side-slip moving on the real vehicle's track is described.

The output of the linear part of the system (2.38) is the input to the nonlinear part (2.39), i.e. the two parts are in serial connection without any feedback branches, which is very convenient for controller design.

# Chapter 3

# Trajectory tracking control

## 3.1 Introduction

The basic task in planar mobile robot motion control is to move to a specified position and orientation, together called posture. Quoting a comprehensive article by Canudas de Wit *et al.* [21], mobile robot motion control is classified into path tracking, trajectory tracking and point stabilisation. A trajectory, in cartesian coordinates given by two functions $x(t)$ and $y(t)$, implicitly contains velocity information, whereas a path is a one-dimensional subset of $\mathbb{R}^2$, i.e. purely geometric. The corresponding orientation angle $\varphi$ is in both cases defined as the tangent to the curve.

Path tracking algorithms rely on previously generated paths, the velocity profile along the path is chosen by the controller. Trajectory tracking algorithms, too, work with previously calculated trajectories, where the velocity profile is already determined in the planning stage. The control problem can be formulated as minimising the error between the actual posture and the momentary reference posture. Most existing tracking algorithms, however, fail at zero reference velocity, i.e. they are incapable of compensating residual posture errors after the reference trajectory reaches a stationary point.

Posture stabilisation does not necessarily involve a pre-planned trajectory or path, it can be defined as asymptotically stabilising the robot at a given (usually static) target posture. Posture stabilisation algorithms can be used for trajectory tracking, if the target posture is not static. Asymptotic posture stabilisation, however, does not guarantee optimal tracking behaviour. A drawback of posture stabilisation without pre-planned trajectory is that there is no possibility to influence the way how the target posture is reached, which makes obstacle avoidance impossible.

A quote from Oriolo [74] casts light on the discrepancy between the different approaches:

> [...] a drawback of many posture stabilizing controllers is a poor transient performance. Another difficulty which has often been overlooked is the necessity of using two different control laws for trajectory tracking and posture stabilization.

Therefore, it is most desirable to develop an algorithm exhibiting both satisfactory tracking and posture stabilisation behaviour. As yet, however, only very few authors have reported the achievement of that goal.

## 3.2 Analysis of the control system

The following analysis is mainly taken from Isidori [50], Murray et al.[70] and Brockett [19].

### 3.2.1 Definitions and classification

The robot's kinematics (2.39) are a classical example for a nonholonomic system, which is typical for a knife edge or a rolling wheel, therefore often called unicycle-type kinematics. The particularity with nonholonomic systems is that the number of global degrees of freedom or states (here: $x$, $y$ and $\theta$) exceeds the number of local degrees of freedom or inputs (here: $v$ and $\omega$) due to a non-integrable kinematic constraint, which is given by

$$-\dot{x}\sin\theta + \dot{y}\cos\theta = 0, \qquad (3.1)$$

equivalent to

$$v_n = 0, \qquad (3.2)$$

i.e. no side-slip is possible. This property causes considerable mathematical complexity, which will become clear throughout the analysis.

The most general form of a control system state space model is given by

$$\dot{\boldsymbol{x}} = \boldsymbol{f}(\boldsymbol{x}, \boldsymbol{u}), \qquad (3.3)$$

where $\boldsymbol{x}$ denotes the state vector and $\boldsymbol{u}$ the input vector, $\boldsymbol{f}(\boldsymbol{x}, \boldsymbol{u})$ is a vector field or nonlinear map $\mathbb{R}^{n+m} \to \mathbb{R}^n$ where $n$ is the dimension of the system, i.e. the number of states, and $m$ is the number of inputs.
A control system which is said to be linear in the inputs can be written as

$$\dot{\boldsymbol{x}} = \boldsymbol{f}(\boldsymbol{x}) + \sum_{i=1}^{m} \boldsymbol{g}_i(\boldsymbol{x}) u_i. \qquad (3.4)$$

Again $\boldsymbol{f}(\boldsymbol{x})$ and $\boldsymbol{g}_i(\boldsymbol{x})$ are vector fields in $\mathbb{R}^n$ and $\boldsymbol{f}(\boldsymbol{x})$ is also called the drift. In a driftless system, $\boldsymbol{f}(\boldsymbol{x})$ vanishes, it is therefore immediately at rest if the input vanishes. Any configuration $\boldsymbol{x}$ represents an equilibrium position in the state space.
The kinematics (2.39) with state vector $\boldsymbol{x} := [x \ y \ \theta]^T$ and inputs $v$ and $\omega$ can be written as

$$\dot{\boldsymbol{x}} = \underbrace{\begin{bmatrix} \cos\theta \\ \sin\theta \\ 0 \end{bmatrix}}_{\boldsymbol{g}_1(\boldsymbol{x})} v + \underbrace{\begin{bmatrix} 0 \\ 0 \\ 1 \end{bmatrix}}_{\boldsymbol{g}_2(\boldsymbol{x})} \omega. \qquad (3.5)$$

and belongs therefore to the class of nonlinear nonholonomic driftless systems, linear in the inputs, dimension $n = 3$ and number of inputs $m = 2$.

## 3.2. ANALYSIS OF THE CONTROL SYSTEM

Because the number of outputs, which are equal to the states, is larger than the number of inputs, this type of systems is sometimes called an *underactuated* system.
For the robot's velocity dynamics (2.38), the drift and the input vector fields are linear in the states, and can therefore be written with state and input matrices. The velocity dynamics are classified as a linear system with two states and two inputs.

### 3.2.2 Controllability

The first important question in the analysis of a control system is its controllability. A system is said to be completely state controllable, if for any configuration $x_0$ in the state space there exists an input $u(t)$ which makes the system reach an arbitrary final configuration $x_f$ in arbitrary time.

A controllability criterion for a system of type (2.39) is presented by Murray *et al.* in [70], based on results by Isidori [50]. This criterion is called the controllability rank condition and is formulated as follows:

**Controllability rank condition** If the controllability Lie algebra has full rank $n$, or rephrased, if the distribution spanned by $g_i$ is not involutive, the system is controllable.

A Lie bracket $[g_1, g_2]$ of two vector fields $g_1$ and $g_2$ is given by

$$[g_1, g_2] = \frac{\partial g_1(x)}{\partial x} g_2(x) - \frac{\partial g_2(x)}{\partial x} g_1(x). \tag{3.6}$$

The controllability Lie algebra is given by

$$\mathcal{L}(g_1, g_2) = (g_1, g_2, [g_1, g_2]) = \begin{bmatrix} \cos\varphi & 0 & -\sin\varphi \\ \sin\varphi & 0 & \cos\varphi \\ 0 & 1 & 0 \end{bmatrix}. \tag{3.7}$$

Its rank is obviously equal to three. Thus the kinematics are controllable.

The controllability of the velocity dynamics can easily be established by checking the rank of the controllability matrix

$$\mathcal{C} = [B, \ AB], \tag{3.8}$$

which clearly has full rank $n = 2$ because $B$ is already a square matrix with full rank.

Since the velocity dynamics and the kinematics are in serial connection, the controllability of the entire system is guaranteed by the controllability of its individual parts.

### 3.2.3 Existence of a smooth stabilising feedback for the kinematics

The controllability only guarantees the existence of a stabilising input. So far, no conclusion can be drawn about the existence of a smooth feedback law or control law $u(x)$.
In fact, Brockett [19] showed, that no smooth time-invariant feedback law exists to stabilise the system (2.39).
Possibilities to circumvent this restriction are therefore:

- Discontinuous control laws.

- Time-varying control laws.

- Hybrid control laws, basically employing different controllers for different domains of the state space.

Trajectory tracking and path tracking algorithms pose less mathematical difficulties, because they are mostly based on the tracking error dynamics of the system, which exhibit more convenient mathematical properties.

### 3.2.4 Decoupling of the velocity dynamics

Decoupling of the states yields two independent first-order single-input-single-output systems, which are very easy to handle. The system matrix $A$ of the velocity dynamics (2.38) is already of diagonal structure, therefore only the input matrix $B$ has to be diagonalised to decouple the two states. An eigenvector-eigenvalue-transformation could be used to diagonalise $B$, but due to its structure a diagonalising transformation is simply found as

$$B = \begin{bmatrix} \dfrac{p_1}{rm_{\text{tot.}}} & \dfrac{p_1}{rm_{\text{tot.}}} \\ \dfrac{p_1 b}{2r I_{\text{tot.}}} & -\dfrac{p_1 b}{2r I_{\text{tot.}}} \end{bmatrix} = \begin{bmatrix} \dfrac{p_1}{rm_{\text{tot.}}} & 0 \\ 0 & \dfrac{p_1 b}{2r I_{\text{tot.}}} \end{bmatrix} \begin{bmatrix} 1 & 1 \\ 1 & -1 \end{bmatrix} = \tilde{B} T, \qquad (3.9)$$

with the non-singular transformation matrix $T$. Then the decoupled system (2.38) reads

$$\dot{v} = A v + \tilde{B} T u = A v + \tilde{B} \tilde{u}, \qquad (3.10)$$

where

$$\tilde{u} = T u = \begin{bmatrix} r_{\text{PWM,R}} + r_{\text{PWM,L}} \\ r_{\text{PWM,R}} - r_{\text{PWM,L}} \end{bmatrix}, \qquad (3.11)$$

i.e. the transformed inputs are the sum of both inputs for the track speed and their difference for the yaw rate, which is a self-evident result.

## 3.3 State of the art

Kim and Tsiotras [59] give an overview of different discontinuous time-invariant and some time-variant controllers, together with an experimental validation on a unicycle-type robot, which, however, seems to operate slow enough to enable the neglection of the velocity dynamics seen from the time-scale of the kinematics, i.e. the velocity dynamics are fast enough compared to the kinematics, that their dynamic response can be considered instantaneous. This property is called *perfect velocity tracking condition*.

Canudas de Wit *et al.* [21] give a very comprehensive overview over the basic analysis and the limitations of the unicycle-type kinematics.
Interesting in this article is the following statement (quoted without proof): The error dynamics of the trajectory tracking problem can be linearised, linear control design is possible. If either $v$ or $\omega$ vanish, controllability is lost.

Another survey paper by Kolmanovsky and McClamroch [60] summarises the various approaches to the unicycle type control problem in tutorial style, including an extensive list of relevant publications.

### 3.3.1 Posture stabilisation by discontinuous control laws

A type of discontinuous control law is sliding mode control: Following a switching law depending on the state, the control input assumes one of two possible extrema.
Theoretically, when the switching frequency is infinitely high, the state trajectory evolves on the switching surface or sliding surface, a manifold of reduced dimension in the state space [94], [25].

Bloch and Drakunov [13] apply sliding mode control to the so-called Brockett-system, or also called Heisenberg-system, a different representation of the unicycle-type kinematics, written as

$$\begin{aligned} \dot{x} &= u \\ \dot{y} &= v \\ \dot{z} &= xv - yu. \end{aligned} \quad (3.12)$$

This representation can be obtained by a state and input transformation given by Hespanha *et al.* in [44].
The article by Bloch and Drakunov [13] is exclusively concerned with the Brockett-system and some generalisations (i.e. the kinematic model), but does not take the velocity dynamics of a real mobile robot into account.

Yang and Kim [102] apply a sliding mode control law to the mobile robot tracking problem written in polar coordinates, taking the velocity dynamics of the robot into account.
However, it appears that there are restrictions on the type of trajectories due to the chosen representation in polar coordinates.

Generally, sliding mode control in mobile robot motion control has the advantage of robustness against modeling uncertainties.

A different type of discontinuous control law is employed by Tanner and Kyriakopoulos in [89], by combining a discontinuous law for the velocity inputs with integrator backstepping, where the original inputs are introduced as additional states and their derivatives are considered as the new inputs.

### 3.3.2 Posture stabilisation and trajectory tracking by time-varying feedback laws

Time-varying feedback laws are feedback laws which explicitly depend on the time variable. Samson [81] and Lee et al. [63] apply time-varying controls both to point stabilisation and trajectory tracking, whereas Divelbiss and Wen [26], Jiang and Nijmeijer [54] and Fliess et al. [37] exclusively concentrate on trajectory tracking, the latter making use of a property called differential flatness of the unicycle-type kinematics.

In the articles of Samson [81] and Jiang and Nijmeijer [54] the unicycle-type kinematics are written in chained form, another canonical form besides the kinematic or cartesian form (2.39) and the Brockett-form (3.12),

$$\begin{aligned} \dot{z}_1 &= v_1 \\ \dot{z}_2 &= v_2 \\ \dot{z}_3 &= z_2 v_1. \end{aligned} \tag{3.13}$$

The transformation to chained form involves a state transformation $z = \Phi(x)$ and a feedback $v = \beta(x)u$, the latter being similar to feedback linearisation procedures, [50]. A derivation and illustrating examples are given by Murray et al. in [70].

In a physical interpretation of the new states $z_1$ denotes the distance along the path, $z_2$ denotes the orientation angle and $z_3$ the lateral distance to the origin, or if the system is written for the deviations from a given path (a simple translation in the state space), $z_3$ denotes the lateral distance to the path.

If the input $v_1$ (the track speed) is considered as a function of time, the problem is reduced to finding stabilising controls for the remaining linear time-variant system, which naturally results in time-varying controls.

On top of that, Samson [81] and Jiang and Nijmeijer [54] derive time-varying controls capable of point stabilisation.

All beforementioned publications concentrate on the nonholonomic kinematics, some of them including systems with a higher number of degrees of freedom, which describe the motion of a car pulling one or more trailers. However, they do not take the velocity dynamics of a nonholonomic vehicle into account, although Jiang and Nijmeijer [54] state that their algorithm could be extendend to such systems.

A somewhat different concept is presented by Oriolo et al. [75]. The beforementioned linear time-variant system is considered as a piecewise linear time-invariant system and the controller coefficients are determined by a cyclic learning algorithm.

### 3.3.3 Posture stabilisation by hybrid control laws and invariant manifold techniques

A hybrid system combines continuous feedback laws with discrete logic. Depending on the state of the system, different continuous, possibly nonlinear, control laws come into effect.

Hespanha and Morse [45, 44] partition the state space in overlapping regions defined by nonlinear functions. The general idea is to keep the states $x_1$ and $x_2$ of the Brockett-system (3.12) large enough to enable the state $x_3$ to converge to zero and subsequently apply a different control law to let $x_1$ and $x_2$ converge to zero. Otherwise, if $x_1$ and $x_2$ were already very small, it would require a considerable control effort to influence $x_3$.

Aguiar and Pascoal [1] introduce the notion of Extended Nonholonomic Double Integrator (ENDI) as opposed to NonHolonomic Integrator (NHI), another synonym for the unicycle-type kinematics, deriving novel state- and input-transformations following the methodology in Isidori [50]. The ENDI-system contains a simplified version of the dynamics of a nonholonomic mobile robot.

Invariant manifold techniques represent a similar approach. An invariant manifold is a hypersurface in the state space with reduced dimension. First, the state of the system is driven onto the invariant manifold, then the origin of the state space is approached on the manifold.
The chained form system (3.13) is subjected to the linear feedback $v_1 = -kz_1$ and $v_2 = -kz_2$, $k$ being a positive constant,

$$\begin{aligned} \dot{z}_1 &= -kz_1 \\ \dot{z}_2 &= -kz_2 \\ \dot{z}_3 &= -kz_1z_2. \end{aligned} \qquad (3.14)$$

This system can be analytically integrated,

$$\begin{aligned} z_1(t) &= z_{10}e^{-kt} \\ z_2(t) &= z_{20}e^{-kt} \\ z_3(t) &= s(\mathbf{z}_0) + \frac{1}{2}z_{10}z_{20}e^{-2kt} \end{aligned} \qquad (3.15)$$

with $s(z) = z_3 - \frac{1}{2}z_1z_2$, $z_{i0}$ being the initial conditions.
The manifold $\mathcal{M} \in \mathbb{R}^2$ is now defined by

$$\mathcal{M} = \{z \in \mathbb{R}^3 : s(\mathbf{z}) = 0\}. \qquad (3.16)$$

If the state trajectory evolves on the manifold $\mathcal{M}$, the origin will be approached exponentially. The manifold $\mathcal{M}$ is invariant for (3.14), because

$$\dot{s} = \dot{z}_3 - \frac{1}{2}(z_1\dot{z}_2 + \dot{z}_1z_2) = -kz_1z_2 + \frac{k}{2}(z_1z_2 + z_1z_2) = 0. \qquad (3.17)$$

Thus, once the state has reached the invariant manifold, it remains there. To reach the manifold, the control law is modified,

$$v_i = -kz_i + \beta_i(\mathbf{z}, s) = -kz_i \pm \mu \frac{s(\mathbf{z})}{z_1^2 + z_2^2}z_i, \quad \mu > 2k, \quad i = 1, 2. \qquad (3.18)$$

This concept by Tsiotras [93] has the drawback of a singularity in the second part of the control law.
Only nonholonomic kinematics are taken into account, the velocities are considered as inputs to the system.
A further development is reported by Luo in [66], using a recursive technique for nonholonomic systems of higher order (e. g. car-trailer-configurations).

It may seem that the proposed control scheme is time-invariant and continuous, which has been proven to be impossible. As all states approach zero, however, the control law exhibits a discontinuity; it is not defined at the origin.

### 3.3.4 Trajectory tracking by feedback linearisation

Static state feedback linearisation is an input and state transformation of the form $z = \Phi(x)$ and $v = \beta(x)u + \alpha(x)$ such that the relationship between $v$ and $z$ is linear.
Full static state feedback linearisation cannot be achieved for the present type of system, as can be shown by a theorem by Isidori [50] (here stated in a somewhat simplified form):

**Theorem (Solvability of the State Space Exact Linearisation Problem for MIMO-Systems)** The State Space Exact Linearisation Problem is solvable if and only if

1. for each $0 \leq i \leq n-2$, the distribution $G_i = \text{span}\{ad_f^k g_j : 0 \leq k \leq i, 1 \leq j \leq m\}$ is involutive;

2. the distribution $G_{n-1}$ has dimension $n$;

where $ad_f^k g(x) = [f, ad_f^{k-1} g(x)]$ and $ad_f^0 g(x) = g(x)$.

Therefore, in the case (2.39), where $f$ vanishes and $n = 3$,

$$G_0 = \text{span}\{g_1, g_2\}$$
$$G_1 = \text{span}\{g_1, g_2, [f, g_1], [f, g_2]\} = \text{span}\{g_1, g_2, 0, 0\} \quad (3.19)$$
$$G_2 = \text{span}\{g_1, g_2, 0, 0, 0, 0\}$$

The first condition is violated, because a distribution spanned by $g_1, g_2$ is not involutive, see section 3.2.2. The second condition is violated, because $G_2$ obviously has only dimension 2.
Note that the very same condition that guarantees the system's controllability prevents it (among others) from being feedback linearisable.
By discarding the orientation angle as output variable and introducing a new state by differentiation, this restriction can be avoided, an approach called dynamic feedback linearisation.

Thorough investigations on the subject of feedback linearisation of nonholonomic systems have been published by d'Andrea-Novel et al. in [23] and previous work, Park et al. in [77] and by De Luca, Oriolo et al. in [24, 74, 22].
The application of the algorithm in [74] is as follows: The output vector is chosen to be $z = [x \ y]^T$, i.e. the orientation angle is omitted. Then, the order of the system is deliberately increased by differentiating once and considering the input $v$ as a new state, a procedure similar

## 3.3. STATE OF THE ART

to integrator backstepping mentioned in Sec. 3.3.1. The desired controllable linear system, where $x$ and $y$ are fully decoupled, is

$$\ddot{z} = v, \qquad (3.20)$$

or alternatively written as linear second order state space system

$$\begin{bmatrix} \dot{x} \\ \ddot{x} \end{bmatrix} = \begin{bmatrix} 0 & 1 \\ 0 & 0 \end{bmatrix} \begin{bmatrix} x \\ \dot{x} \end{bmatrix} + \begin{bmatrix} 0 \\ 1 \end{bmatrix} v_1, \qquad (3.21)$$

and analogously for $y$, introducing a modified input $v$, which is given by

$$v = \underbrace{\begin{bmatrix} \cos\theta & -v\sin\theta \\ \sin\theta & v\cos\theta \end{bmatrix}}_{T(\theta,v)} \begin{bmatrix} \dot{v} \\ \omega \end{bmatrix}. \qquad (3.22)$$

For the modified input $v$ a linear control law is designed applying standard state-feedback design procedures to the linear controllable single-input-single-output systems (3.21). To obtain the actual control inputs $v$ and $\omega$, the transformation matrix $T(\theta, v)$ is inverted, and $\dot{v}$ is integrated. To preserve full rank of $T(\theta, v)$ the velocity input $v$ must not become zero, a restriction common to most trajectory tracking schemes.
Again, the velocity dynamics are not considered.

### 3.3.5 Trajectory tracking via a nonlinear control law for the error dynamics

Based on a representation of the posture error dynamics in a local (fixed to the robot) frame, a simple nonlinear control scheme was first developed by Kanayama et al. in [57].
The posture errors in local coordinates are given by

$$\begin{bmatrix} e_1 \\ e_2 \\ e_3 \end{bmatrix} = \begin{bmatrix} \cos\theta & \sin\theta & 0 \\ -\sin\theta & \cos\theta & 0 \\ 0 & 0 & 1 \end{bmatrix} \begin{bmatrix} x_r - x \\ y_r - y \\ \theta_r - \theta \end{bmatrix}, \qquad (3.23)$$

where the index r denotes the respective reference quantities. The derivatives of the errors in the local frame are then given by

$$\dot{e} = \begin{bmatrix} \omega e_2 - v + v_r \cos e_3 \\ -\omega e_1 + v_r \sin e_3 \\ \omega_r - \omega \end{bmatrix}. \qquad (3.24)$$

Kanayama et al. [57] give a smooth velocity input that lets the posture error $e$ converge to zero, i.e. enables trajectory tracking,

$$\begin{bmatrix} v \\ \omega \end{bmatrix} = \begin{bmatrix} v_r \cos e_3 + k_1 e_1 \\ \omega_r + k_2 v_r e_2 + k_3 v_r \sin e_3 \end{bmatrix}. \qquad (3.25)$$

Convergence and stability is guaranteed using Lyapunov-analysis, as long as $v_r, \omega_r \neq 0$.
Fierro and Lewis extend Kanayamas result by including the velocity dynamics through integrator backstepping in [36], Hu and Yang in [46] include uncertain dynamics and Jiang and Nijmeijer apply integrator backstepping to a slightly different control law in [53].

## 3.3.6 Path tracking

Only very few publications deal with pure path tracking, among them are Wu *et al.* [101], employing integrator backstepping to a representation of the nonholonomic kinematics in polar coordinates, Breivik and Fossen [18] and Egerstedt *et al.* [28].

## 3.4 Introduction to nonlinear predictive control

In nonlinear predictive control a sequence of optimal future control inputs is iteratively calculated during every sampling interval using a suitable numerical optimisation algorithm and working with some discrete dynamic model of the plant. A scalar cost function weighing predicted future control errors against future control inputs up to a so-called prediction horizon is minimised during every sampling interval. The principle of nonlinear predictive control is adapted from Norgaard [71], where a similar procedure is applied to single-input-single-output systems, which are dynamically modeled by recurrent multilayer perceptron networks.

In this work, predictive control is used in conjunction with linear state feedback control, representing a form of cascaded control scheme, see Fig. 3.1. Minimisation is performed using a Gauss-Newton algorithm. The velocities (i.e. track speed $v$ and yaw rate $\omega$) are controlled by a linear state feedback control law, representing the cascade's inner loop, whereas the posture $x$, $y$ and $\varphi$ is controlled by predictive control in the outer loop, where the prediction makes use of a linear closed-loop representation of the inner loop. The choice of a cascaded control scheme enables the direct incorporation of velocity feedback, whereas a predictive control scheme for the entire system would only use the measured posture for feedback.

The reference velocities of the inner loop serve as the control inputs in the outer loop, i.e. the predictive controller calculates a sequence of reference velocities to minimise the predicted future deviation from the pre-planned trajectory under consideration of the closed-loop dynamics of the velocity control loop. This implies that future reference postures up to the prediction horizon must be known.

Since the optimisation is performed during every sampling interval, only the first values of the calculated sequence of reference velocities are ever actually applied to the system. The velocity control law is then used to calculate the corresponding PWM-input to the DC-motors.

The algorithm is augmented by integration of the position errors in the outer loop and an additional measure for the compensation of lateral position errors, which enables posture stabilisation on top of trajectory tracking. Moreover, a possible side-slip angle is explicitly accounted for.

The application of nonlinear predictive control to a nonholonomic control problem constitutes a radically different way to circumvent the beforementioned control restrictions, since it does not include any explicit control law. The controllability of the system, however, guarantees a unique solution to the optimisation procedure.

The explicit incorporation of the side-slip angle in the control algorithm is a novelty and as yet no similar approach to the problem of lateral wheel slip seems to exist, at least not throughout the accessible literature on mobile robot motion control.

A preliminary version of the algorithm presented in this chapter is also documented in a conference contribution to the 2005 International Conference on Informatics in Control, Automation and Robotics [82] and in the current form in a journal article in Robotics and Autonomous

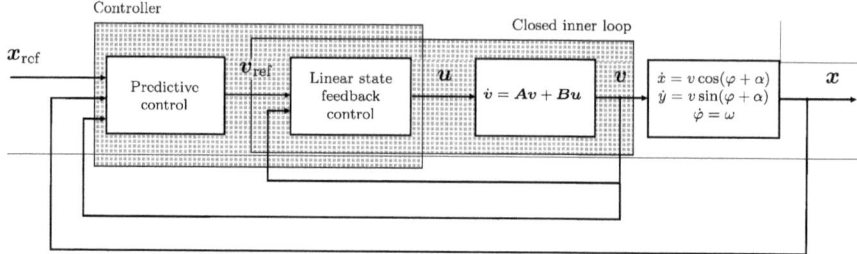

Figure 3.1: Cascaded control scheme.

Systems [52].

The results presented in a conference contribution to the 2005 16th IFAC world congress [85] concerning neural network predictive control of a mobile robot are omitted in this work.

## 3.5 Linear control of the inner loop

### 3.5.1 Discretisation of the linear velocity dynamics

First, the continuous-time state space representation of the robot's dynamics is discretised for the sampling time of the control algorithm $T_C = 0.02$s assuming that a zero-order hold acts at the input. Theoretical background is presented e.g. by Ogata in [73].
The decoupled system (3.9) is then written as

$$\boldsymbol{v}_{K+1} = \underbrace{\begin{bmatrix} a_1 & 0 \\ 0 & a_2 \end{bmatrix}}_{A} \boldsymbol{v}_K + \underbrace{\begin{bmatrix} b_1 & 0 \\ 0 & b_2 \end{bmatrix}}_{B} \underbrace{\boldsymbol{T}\boldsymbol{u}_K}_{\tilde{\boldsymbol{u}}_K}, \tag{3.26}$$

where $a_1, a_2, b_1, b_2$ are real-valued constants and $K$ denotes the integer sampling instant.

### 3.5.2 Linear control law

A linear state feedback control law as shown in Fig. 3.2, of the form

$$\begin{aligned} \tilde{u}_{1,K} &= k_{w1} v_{\text{ref},K} - k_1 v_K \\ \tilde{u}_{2,K} &= k_{w2} \omega_{\text{ref},K} - k_2 \omega_K, \end{aligned} \tag{3.27}$$

where the index ref denotes the reference quantities, is found by pole assignment and adjustment of the stationary gain. Inserting (3.27) into (3.26) yields

$$\begin{aligned} v_{K+1} &= \underbrace{(a_1 - b_1 k_1)}_{p_1} v_K + \underbrace{b_1 k_{w1}}_{1-p_1} v_{\text{ref},K} \\ \omega_{K+1} &= \underbrace{(a_2 - b_2 k_2)}_{p_2} \omega_K + \underbrace{b_2 k_{w2}}_{1-p_2} \omega_{\text{ref},K}. \end{aligned} \tag{3.28}$$

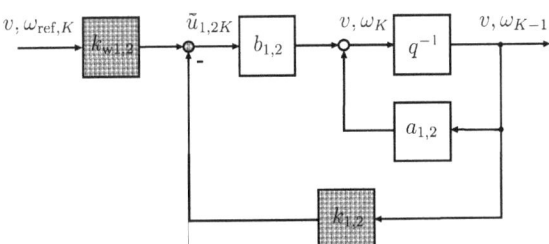

Figure 3.2: Linear control of the inner loop, grey areas symbolise the controller. $q^{-1}$ denotes a time-shift.

The feedback gains $k_1$ and $k_2$ are chosen such that $a_1 - b_1 k_1$ and $a_2 - b_2 k_2$ are equal to the desired poles $p_1$ and $p_2$, respectively. The pre-gains are calculated according to $k_{w1} = (1-p_1)/b_1$ and $k_{w2} = (1-p_2)/b_2$, thereby the stationary gain of the closed-loop system is equal to one.

Finally, from $\tilde{\boldsymbol{u}}$ the original inputs $\boldsymbol{u}$ are recovered according to $\boldsymbol{u} = \boldsymbol{T}^{-1}\tilde{\boldsymbol{u}}$, thus the feedback matrix

$$\boldsymbol{K} = \boldsymbol{T}^{-1} \begin{bmatrix} k_1 & 0 \\ 0 & k_2 \end{bmatrix} \in \mathbb{R}^{2\times 2} \tag{3.29}$$

and the pre-gain matrix

$$\boldsymbol{K}_{\mathrm{w}} = \boldsymbol{T}^{-1} \begin{bmatrix} k_{\mathrm{w}1} & 0 \\ 0 & k_{\mathrm{w}2} \end{bmatrix} \in \mathbb{R}^{2\times 2} \tag{3.30}$$

are obtained to form the control law

$$\boldsymbol{u}_K = \boldsymbol{K}_{\mathrm{w}} \underbrace{\begin{bmatrix} v_{\mathrm{ref},K} \\ \omega_{\mathrm{ref},K} \end{bmatrix}}_{\boldsymbol{v}_{\mathrm{ref},K}} - \boldsymbol{K}\boldsymbol{v}_K. \tag{3.31}$$

## 3.6 Nonlinear predictive control of the outer loop

### 3.6.1 Discretisation of the nonlinear kinematic model

A simple forward difference approximation (forward Euler scheme) for the first order derivatives is applied, analogously for $y$ and $\varphi$,

$$\dot{x}_K = \frac{x_{K+1} - x_K}{T_C}, \tag{3.32}$$

$$\begin{aligned} x_{K+1} &= T_C\, v_K \cos(\varphi_K + \alpha_K) + x_K \\ y_{K+1} &= T_C\, v_K \sin(\varphi_K + \alpha_K) + y_K \\ \varphi_{K+1} &= T_C\, \omega_K + \varphi_K. \end{aligned} \tag{3.33}$$

### 3.6.2 Cost function

The proposed scalar quadratic cost function contains predicted future position errors $\boldsymbol{P}$ up to the prediction horizon $h_\mathrm{p}$, their cumulative sum $\boldsymbol{Q}$ which corresponds to a discrete integration, and the control variable increments $\Delta \boldsymbol{v}_\mathrm{ref}$.

In digital control applications it is usually tacitly assumed that the calculation time of the control algorithm is negligible. Since the presented algorithm, however, must be expected to consume considerable calculation time, this assumption does not hold. On the contrary, it is assumed that the calculated control input can only be applied at the next sampling instant, such that the algorithm can take a full sampling interval to complete the calculations.

Therefore, and because the combination of the dynamics of the inner loop and the kinematics is of second order, the earliest outputs that can be influenced by the reference velocities calculated between instant $K$ and instant $K+1$ are those at instant $K+3$.

The reference velocities up to $K + h_\mathrm{p} - 2$ are assumed to be constant. For the initial cycle of the iteration they are chosen to be equal to those calculated between $K-1$ and $K$. The increment $\Delta \boldsymbol{v}_\mathrm{ref}$ therefore denotes the difference of reference velocities between two subsequent cycles of the iteration. The choice of the reference velocity increment instead of its absolute value aims at achieving a smooth reference velocity input, i.e. at controlling its bandwidth.

The scalar, positive definite cost function to be minimised reads

$$V = \frac{1}{2}\Delta \boldsymbol{v}_\mathrm{ref}^\mathrm{T} \boldsymbol{R} \Delta \boldsymbol{v}_\mathrm{ref} + \frac{1}{2 h_\mathrm{p}} \boldsymbol{P}^\mathrm{T} \boldsymbol{L}_P \boldsymbol{P} + \frac{1}{2} \boldsymbol{Q}^\mathrm{T} \boldsymbol{T}^\mathrm{T} \boldsymbol{L}_Q \boldsymbol{T} \boldsymbol{Q} \to \min_{\Delta \boldsymbol{v}_\mathrm{ref}}. \tag{3.34}$$

The predicted posture errors from instant $K+3$ up to instant $K + h_\mathrm{p}$ are concatenated in one vector,

$$P = \begin{bmatrix} x_{\text{ref},K+3} - \hat{x}_{K+3} \\ \vdots \\ x_{\text{ref},K+h_\text{p}} - \hat{x}_{K+h_\text{p}} \\ y_{\text{ref},K+3} - \hat{y}_{K+3} \\ \vdots \\ y_{\text{ref},K+h_\text{p}} - \hat{y}_{K+h_\text{p}} \\ \varphi_{\text{ref},K+3} - \hat{\varphi}_{K+3} \\ \vdots \\ \varphi_{\text{ref},K+h_\text{p}} - \hat{\varphi}_{K+h_\text{p}} \end{bmatrix}. \tag{3.35}$$

The integrated predicted position errors are given by

$$Q = q_K + \begin{bmatrix} \sum_{i=1}^{h_\text{p}} x_{\text{ref},K+i} - \hat{x}_{K+i} \\ \sum_{i=1}^{h_\text{p}} y_{\text{ref},K+i} - \hat{y}_{K+i} \end{bmatrix}, \tag{3.36}$$

where

$$q_K = q_{K-1} + \begin{bmatrix} x_{\text{ref},K} - x_K \\ y_{\text{ref},K} - y_K \end{bmatrix}. \tag{3.37}$$

The true position errors are integrated in $q$, whereas on top of that during every sampling interval the predicted control errors are integrated in $Q$ but limited to a certain interval to avoid unwanted windup.

The transformation matrix

$$T = \begin{bmatrix} \cos \varphi_{\text{ref},K} & \sin \varphi_{\text{ref},K} \\ -\sin \varphi_{\text{ref},K} & \cos \varphi_{\text{ref},K} \end{bmatrix} \tag{3.38}$$

transforms the integrated predicted position errors into local coordinates, such that they can be individually weighted.

The weight matrices $L_P \in \mathbb{R}^{3(h_\text{p}-2) \times 3(h_\text{p}-2)}$, $L_Q \in \mathbb{R}^{2 \times 2}$ and $R \in \mathbb{R}^{2 \times 2}$ determine the trade-off between future control errors, control error integrals and the bandwidth of the reference velocity inputs.

### 3.6.3 Prediction

The velocities are predicted using the closed loop dynamic state space representation of the inner loop,

$$v_{K+1} = \underbrace{(A - BK)}_{\hat{A}} v_K + \underbrace{BK_\text{w}}_{\hat{B}} v_{\text{ref},K}, \tag{3.39}$$

where both the closed-loop system matrix $\hat{A}$ and the closed-loop input matrix $\hat{B}$ have diagonal structure, i.e. the states are fully decoupled.

## 3.6. NONLINEAR PREDICTIVE CONTROL OF THE OUTER LOOP

Estimates of the future velocities are recursively calculated up to instant $K + h_p - 1$. The reference velocities $v_{\text{ref},K}$ are those calculated between $K - 1$ and $K$, whereas $v_{\text{ref},K+1}$ through $v_{\text{ref},K+h_p-2}$ are updated after every cycle of the iterative optimisation procedure.
The positions are predicted by using (3.33) under the simple assumption that the side-slip angle $\alpha$ remains constant up to the prediction horizon.

### 3.6.4 Incorporation of the side-slip angle

A curved path requires centripetal acceleration, which can only be provided by a lateral force transmitted by the wheels. Due to the wheels' friction characteristic, a lateral force is a direct consequence of a certain side-slip velocity. Thus a certain side-slip angle is inevitable while moving along a curve. It is therefore impossible to keep the robot's orientation tangential to the curve and at the same time keep its center of gravity on the curved path. The reference orientation angle must be modified by subtracting a large fraction $r_\alpha$ of the current side-slip angle $\alpha_K$ from the reference path angle $\theta_{\text{ref},K}$,

$$\varphi_{\text{ref},K} = \theta_{\text{ref},K} - \alpha_K r_\alpha, \quad r_\alpha < 1. \tag{3.40}$$

This can be interpreted as the two-wheeled equivalent of power-drifting of four-wheeled vehicles.

### 3.6.5 Compensation of lateral position errors

To enable compensation of a lateral position error, the attitude angle is further modified. The reference attitude is rotated by a small angle $\psi$ towards the reference position, see Fig. 3.3. Otherwise a lateral offset from the reference position represents a local minimum of the cost function (3.34). To reduce the position error, a temporary increase in attitude error would be necessary, which would in turn increase the value of the cost function.
The final reference attitude is modified according to

$$\varphi_{\text{ref},K} + \psi \, \text{sign}(p_{\text{n},K}) \, \text{sign}(v_K) \, \exp(-a|v_K|), \quad a > 0, \tag{3.41}$$

where

$$p_{\text{n},K} = -\sin \tilde{\varphi}_{\text{ref},K} + \cos \tilde{\varphi}_{\text{ref},K} \tag{3.42}$$

denotes the lateral position error. The sign of $v_K$ adjusts the direction of the adaptation according to the direction of motion and the exponential function reduces the effect at higher track speeds.

### 3.6.6 Minimisation

The posture errors $\boldsymbol{P}$ are approximated by a first order Taylor series expansion,

$$\boldsymbol{P} \doteq \boldsymbol{P}_0 + \frac{\partial \boldsymbol{P}}{\partial \boldsymbol{v}_{\text{ref}}} \Delta \boldsymbol{v}_{\text{ref}} = \boldsymbol{P}_0 - \frac{\partial \boldsymbol{X}}{\partial \boldsymbol{v}_{\text{ref}}} \Delta \boldsymbol{v}_{\text{ref}} := \boldsymbol{P}_0 - \boldsymbol{D}_P \Delta \boldsymbol{v}_{\text{ref}} \tag{3.43}$$

The matrix $\boldsymbol{D}_P \in \mathbb{R}^{3(h_p-2) \times 2}$ is given by

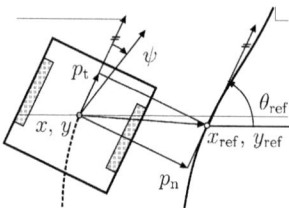

Figure 3.3: Modification of the reference attitude to enable compensation of lateral position errors.

$$\boldsymbol{D}_P = \begin{bmatrix} \dfrac{\partial x_{K+2+i}}{\partial v_{\mathrm{ref},K+j}} \underbrace{\dfrac{\partial v_{\mathrm{ref},K+j}}{\partial \Delta v_{\mathrm{ref}}}}_{1} & \dfrac{\partial x_{K+2+i}}{\partial \omega_{\mathrm{ref},K+j}} \underbrace{\dfrac{\partial \omega_{\mathrm{ref},K+j}}{\partial \Delta \omega_{\mathrm{ref}}}}_{1} \\ \dfrac{\partial y_{K+2+i}}{\partial v_{\mathrm{ref},K+j}} \underbrace{\dfrac{\partial v_{\mathrm{ref},K+j}}{\partial \Delta v_{\mathrm{ref}}}}_{1} & \dfrac{\partial y_{K+2+i}}{\partial \omega_{\mathrm{ref},K+j}} \underbrace{\dfrac{\partial \omega_{\mathrm{ref},K+j}}{\partial \Delta \omega_{\mathrm{ref}}}}_{1} \\ \dfrac{\partial \varphi_{K+2+i}}{\partial v_{\mathrm{ref},K+j}} \underbrace{\dfrac{\partial v_{\mathrm{ref},K+j}}{\partial \Delta v_{\mathrm{ref}}}}_{1} & \dfrac{\partial \varphi_{K+2+i}}{\partial \omega_{\mathrm{ref},K+j}} \underbrace{\dfrac{\partial \omega_{\mathrm{ref},K+j}}{\partial \Delta \omega_{\mathrm{ref}}}}_{1} \end{bmatrix} = \begin{bmatrix} \sum_{j=1}^{h_{\mathrm{p}}-2} \dfrac{\partial x_{K+2+i}}{\partial v_{\mathrm{ref},K+j}} & \sum_{j=1}^{h_{\mathrm{p}}-2} \dfrac{\partial x_{K+2+i}}{\partial \omega_{\mathrm{ref},K+j}} \\ \sum_{j=1}^{h_{\mathrm{p}}-2} \dfrac{\partial y_{K+2+i}}{\partial v_{\mathrm{ref},K+j}} & \sum_{j=1}^{h_{\mathrm{p}}-2} \dfrac{\partial y_{K+2+i}}{\partial \omega_{\mathrm{ref},K+j}} \\ \sum_{j=1}^{h_{\mathrm{p}}-2} \dfrac{\partial \varphi_{K+2+i}}{\partial v_{\mathrm{ref},K+j}} & \sum_{j=1}^{h_{\mathrm{p}}-2} \dfrac{\partial \varphi_{K+2+i}}{\partial \omega_{\mathrm{ref},K+j}} \end{bmatrix} \quad (3.44)$$

with $i, j \in [1; h_{\mathrm{p}} - 2]$ and $\mathbf{1} \in \mathbb{R}^{h_{\mathrm{p}}-2 \times 1}$.
The position error integrals $\boldsymbol{Q}$ are also linearised with respect to the reference velocities, much in the same way as the posture errors,

$$\boldsymbol{Q} \doteq \boldsymbol{Q}_0 + \frac{\partial \boldsymbol{Q}}{\partial \boldsymbol{v}_{\mathrm{ref}}} \Delta \boldsymbol{v}_{\mathrm{ref}} := \boldsymbol{Q}_0 - \boldsymbol{D}_Q \Delta \boldsymbol{v}_{\mathrm{ref}}, \quad (3.45)$$

and since $\boldsymbol{Q}$ contains the sum of the elements of $\boldsymbol{P}$, $\boldsymbol{D}_Q \in \mathbb{R}^{2 \times 2}$ is given by

$$\boldsymbol{D}_Q = \begin{bmatrix} \sum_{j=1}^{h_{\mathrm{p}}-2} \sum_{i=1}^{h_{\mathrm{p}}-2} \dfrac{\partial x_{K+2+i}}{\partial v_{\mathrm{ref},K+j}} & \sum_{j=1}^{h_{\mathrm{p}}-2} \sum_{i=1}^{h_{\mathrm{p}}-2} \dfrac{\partial x_{K+2+i}}{\partial \omega_{\mathrm{ref},K+j}} \\ \sum_{j=1}^{h_{\mathrm{p}}-2} \sum_{i=1}^{h_{\mathrm{p}}-2} \dfrac{\partial y_{K+2+i}}{\partial v_{\mathrm{ref},K+j}} & \sum_{j=1}^{h_{\mathrm{p}}-2} \sum_{i=1}^{h_{\mathrm{p}}-2} \dfrac{\partial y_{K+2+i}}{\partial \omega_{\mathrm{ref},K+j}} \end{bmatrix} \quad (3.46)$$

The derivatives of the positions with respect to the reference velocities are calculated recursively, here exemplarily conducted for $x$ and $v_{\mathrm{ref}}$. From (3.33)

$$x_{K+2+i} = T_{\mathrm{C}} v_{K+1+i} \cos(\varphi_{K+1+i} + \alpha_K) + x_{K+1+i} \quad (3.47)$$

is found and from (3.39)

$$v_{K+1+i} = \hat{A}_{11} v_{K+i} + \hat{B}_{11} v_{\mathrm{ref}\,K+i}. \quad (3.48)$$

Three different cases have to be distinguished:

## 3.6. NONLINEAR PREDICTIVE CONTROL OF THE OUTER LOOP

1. If $i = j$, i.e. the direct influence of the reference velocity $v_{\text{ref}}$ on the coordinate $x$ two sampling instants later is calculated, the derivative is directly given by

$$\frac{\partial x_{K+2+i}}{\partial v_{\text{ref}K+j}} = \frac{\partial x_{K+2+i}}{\partial v_{K+1+i}} \frac{\partial v_{K+1+i}}{\partial v_{\text{ref}K+j}} = T_C \cos(\varphi_{K+1+i} + \alpha_K) \hat{B}_{11}. \quad (3.49)$$

2. If $i > j$, i.e. the indirect influence of the reference velocity on the coordinate more than two sampling instants later is sought, the derivative includes previously calculated derivatives,

$$\frac{\partial x_{K+2+i}}{\partial v_{\text{ref}K+j}} = \frac{\partial x_{K+2+i}}{\partial v_{K+1+i}} \frac{\partial v_{K+1+i}}{\partial v_{\text{ref}K+j}} + \frac{\partial x_{K+2+i}}{\partial \varphi_{K+1+i}} \underbrace{\frac{\partial \varphi_{K+1+i}}{\partial v_{\text{ref}K+j}}}_{\text{prev.}} + \frac{\partial x_{K+2+i}}{\partial x_{K+1+i}} \underbrace{\frac{\partial x_{K+1+i}}{\partial v_{\text{ref}K+j}}}_{\text{prev.}} =$$

$$= T_C \cos(\varphi_{K+1+i} + \alpha_K) \hat{B}_{11} \hat{A}_{11}^{i-j} - v_{K+1+i} T_C \sin(\varphi_{K+1+i} + \alpha_K) \underbrace{\frac{\partial \varphi_{K+1+i}}{\partial v_{\text{ref}K+j}}}_{\text{prev.}} + 1 \cdot \underbrace{\frac{\partial x_{K+1+i}}{\partial v_{\text{ref}K+j}}}_{\text{prev.}}$$

$$(3.50)$$

3. If $i < j$ obviously all derivatives are equal to zero.

Inserting (3.43) and (3.45) into (3.34) yields

$$V(\Delta \boldsymbol{v}_{\text{ref}}) = \frac{1}{2} \Delta \boldsymbol{v}_{\text{ref}}^{\text{T}} \boldsymbol{R} \Delta \boldsymbol{v}_{\text{ref}} + \frac{1}{2h_{\text{p}}} (\boldsymbol{P}_0 - \boldsymbol{D}_P \Delta \boldsymbol{v}_{\text{ref}})^{\text{T}} \boldsymbol{L}_P (\boldsymbol{P}_0 - \boldsymbol{D}_P \Delta \boldsymbol{v}_{\text{ref}}) +$$
$$+ \frac{1}{2} (\boldsymbol{Q}_0 - \boldsymbol{D}_Q \Delta \boldsymbol{v}_{\text{ref}})^{\text{T}} \boldsymbol{T}^{\text{T}} \boldsymbol{L}_Q \boldsymbol{T} (\boldsymbol{Q}_0 - \boldsymbol{D}_Q \Delta \boldsymbol{v}_{\text{ref}}). \quad (3.51)$$

Differentiation of (3.51) with respect to $\Delta \boldsymbol{v}_{\text{ref}}$, setting the derivative equal to zero and some algebraic manipulations yield the linear system of equations with dimension two

$$\left( \boldsymbol{R} + \frac{1}{h_{\text{p}}} \boldsymbol{D}_P^{(\nu)\text{T}} \boldsymbol{L}_P \boldsymbol{D}_P^{(\nu)} + \boldsymbol{D}_Q^{(\nu)\text{T}} \boldsymbol{T}^{\text{T}} \boldsymbol{L}_Q \boldsymbol{T} \boldsymbol{D}_Q^{(\nu)} \right) \Delta \boldsymbol{v}_{\text{ref}}^{(\nu)} = -\frac{1}{h_{\text{p}}} \boldsymbol{D}_P^{(\nu)\text{T}} \boldsymbol{L}_P \boldsymbol{P}_0^{(\nu)} - \boldsymbol{D}_Q^{(\nu)\text{T}} \boldsymbol{T}^{\text{T}} \boldsymbol{L}_Q \boldsymbol{T} \boldsymbol{Q}_0^{(\nu)}$$

$$(3.52)$$

written with index $\nu$ for the $\nu$-th cycle of the iteration, which can easily be solved for $\Delta \boldsymbol{v}_{\text{ref}}^{(\nu)}$. The reference velocities are updated according to

$$\underbrace{\boldsymbol{v}_{\text{ref}}^{(\nu+1)}}_{[2(h_{\text{p}}-1) \times 1]} = \underbrace{\boldsymbol{v}_{\text{ref}}^{(\nu)}}_{[2(h_{\text{p}}-1) \times 1]} + \underbrace{\begin{bmatrix} 0 & 0 \\ 1 & 0 \\ 0 & 0 \\ 0 & 1 \end{bmatrix}}_{[2(h_{\text{p}}-1) \times 2]} \underbrace{\Delta \boldsymbol{v}_{\text{ref}}^{(\nu)}}_{[2 \times 1]} \quad (3.53)$$

where the initial estimate $\boldsymbol{v}_{\text{ref}}^{(0)}$ is given by the reference velocities of the previous sampling instant, as mentioned before. Then, the updated prediction of the position errors $\boldsymbol{P}_0^{(\nu+1)}$, their cumulative sum $\boldsymbol{Q}_0^{(\nu+1)}$ and the matrices of derivatives $\boldsymbol{D}_{P,Q}^{(\nu+1)}$ at the new predicted positions are calculated.

After a specified number of iterations, the algorithm terminates. Only a few cycles are sufficient to achieve convergence.

The design parameters are:

- The (diagonal) weight matrices $\boldsymbol{L}_{P,Q}$ and $\boldsymbol{R}$: Determine the trade-off between accuracy and bandwidth of the system.

- Prediction horizon $h_\mathrm{p}$: Chosen as small as possible to save calculation time but large enough to ensure smooth and stable tracking.

- Sampling time $T_\mathrm{C}$: Mainly depending on the hardware. Criteria are time constants of the robot's dynamics and its computational abilities.

- Relaxation parameter $r_\alpha$: Determines to what extent a possible side-slip angle is considered. Chosen as close to one as possible, but small enough to avoid unwanted oversteering.

- Adaptation angle $\psi$: Chosen large enough to enable compensation of lateral position errors, but as small as possible.

- Number of iterations performed: Large enough to achieve convergence, but as small as possible to save calculation time.

## 3.6.7 Calculation of the control inputs

Finally, the PWM-control inputs to the DC-motors are calculated using the control law (3.31) at instant $K+1$. Therefore, the optimised reference velocities $\boldsymbol{v}_{\mathrm{ref},K+1}$ and the predicted velocities $\boldsymbol{v}_{K+1}$ are used in

$$\boldsymbol{u}_{K+1} = \boldsymbol{K}_\mathrm{w}\boldsymbol{v}_{\mathrm{ref},K+1} - \boldsymbol{K}\boldsymbol{v}_{K+1}. \qquad (3.54)$$

At instant $K+1$ the control inputs $\boldsymbol{u}_{K+1}$ are applied to the DC-motors via the PWM-module of the microcontroller and held constant for one sampling interval.

To compensate the discrepancy between the nonlinear characteristic and the linear approximation as shown in Fig. 2.13, at low PWM-values, which can be described as a dead-zone, any control input with an absolute value below 0.04 but unequal to zero is replaced by 0.04 sign $r_\mathrm{PWM}$.

# Chapter 4

# Navigation

## 4.1 Introduction

In planar mobile robot motion control navigation is defined as the estimation of the current posture relative to some world reference frame. According to the type of employed sensors, navigation can be classified into proprioceptive navigation, also called dead reckoning, and exteroceptive navigation, as defined by Fabrizi *et al.* in [32]. Usually it is necessary to combine sensor data from different types of sensors to obtain reliable posture information, therefore navigation is often called a sensor fusion problem.

Proprioceptive navigation obtains a posture estimate through continuous update according to a kinematic model of the robot. Examples for proprioceptive sensors are acceleration sensors or wheel angular velocity sensors. If the continuity of the sensor information is compromised, the posture estimation fails and cannot be recovered. The posture estimate contains the entire time-history of measurement errors, and is therefore invariably subject to unbounded accumulation, i.e. eventually the vehicle will entirely lose track of its position.

Proprioceptive navigation is further subdivided: Navigation relying on the measurement of wheel revolutions is called odometry whereas navigation based on the measurement of accelerations and angular velocities relative to the world reference frame is called inertial navigation.

Exteroceptive navigation relies on instantaneous perception of the surrounding environment, either by active devices such as GPS satellites and radio beacons, which strictly speaking violates the conditions for autonomous operation, or passively by vision systems, ultrasonic sensing or laser scans, combined with some map of the environment. The posture error is only compromised by a bounded uncertainty as imposed by the sensor's mode of operation and the uncertainty of the map, but does not accumulate.

Even though their posture estimate is eventually subject to unbounded error, it is still desirable to improve proprioceptive navigation systems, since they increase the allowable travel distance between absolute position updates, as stated by Borenstein and Feng in [15].

## 4.2 State of the art

The standard method for sensor fusion is the extended Kalman filter (EKF), which goes back to a famous article by Kalman in 1960, [55]. An EKF provides an optimal estimate in the sense that the statistical variance of the estimated states is minimised. In most publications concerned with mobile robot navigation, the EKF is employed.

Fuke and Krotkov use an EKF for sensor fusion in [41] for three-dimensional navigation of a lunar rover travelling over uneven terrain. Odometric data from wheel angular velocity sensors and inertial data from acceleration sensors and gyroscopic sensors is fused. Von der Hardt *et al.* combine odometry, compass and gyroscopic sensors via EKF in [97], introducing a method for the compensation of magnetic field influences on the compass.

A number of publications are concerned with a combination of proprioceptive and exteroceptive navigation. Roumeliotis and Bekey [80] use an EKF to combine dead reckoning with sparse exteroceptive data from a sensor measuring the relative position of the sun in the navigation concept of an experimental Mars rover. Fabrizi *et al.* [33] feed inertial sensor data and ultrasonic measurement to an EKF.

In [32] Fabrizi *et al.* enhance the uncertainty modeling of an EKF, which is usually done by a constant state covariance matrix, by variable state covariances, depending on the linear and angular velocities and employing fuzzy rules.

None of these publications, however, addresses a fundamental problem in state estimation via EKF: An EKF aims at the reduction of the statistical variance of the estimated states, but it is implicitly assumed that the measurements are free of bias. Inertial sensors, however, especially acceleration sensors, are strongly affected by bias errors, which can also drift over time or depend on temperature. Moreover, since proprioceptive navigation relies on a kinematic model, an error is possibly introduced by inaccurate geometric parameters. According to a definition by Borenstein and Feng in [16], both types of errors belong to the class of systematic errors.

Different approaches have been reported to deal with the problem of systematic measurement errors. Solda *et al.* [86] combine odometric and gyroscopic data, where the gyroscope's bias is continuously estimated, thus increasing accuracy. Barshan and Durrant-Whyte [10] develop error models, which are generated offline, and incorporated in a three-dimensional EKF for inertial navigation. Borenstein and Feng [16] develop a statistical method to calibrate wheel diameter and wheelbase, which are vital to odometric navigation.

Non-systematic errors are introduced into odometry by discontinuous ground contact or wheel slip. Borenstein and Feng [15] therefore combine odometric and gyroscopic data in a different way, called gyrodometry: Most of the time, odometry is used exclusively, while gyro data is substituted only during the brief instances during which gyro and odometry data differ substantially. Mazl and Preucil [67] present a yet more comprehensive concept for train localisation, where GPS, inertial data and odometric data is used for mutual calibration and data validation. They introduce the term rule-based substitution, which also describes the concept by Borenstein and Feng very well.

## 4.3. AVAILABLE SENSORS

A very comprehensive overview of sensors and methods for mobile robot navigation by Borenstein et al. is found in [14].

## 4.3 Available sensors

At the current stage of development, the robot is equipped with two single-chip two-axis acceleration sensors measuring tangential and lateral acceleration, a single-chip gyro sensor and incremental shaft encoders for each side, see Fig. 4.1. Thus navigation is limited to purely proprioceptive navigation.

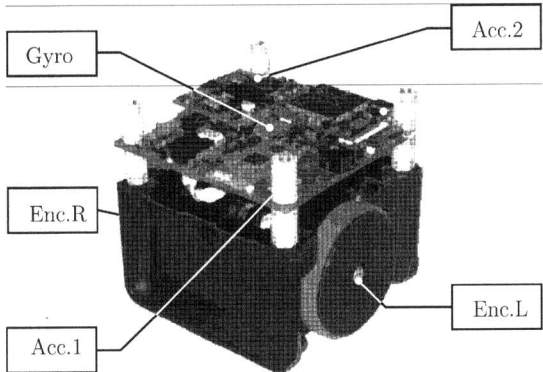

Figure 4.1: Tinyphoon's on-board sensors.

The sensors' characteristics such as noise level and measurement range are compiled in Appendix A.

### 4.3.1 Calibration of the accelerometers

Under the assumption of a linear sensor characteristic, calibration is performed by determining bias $b$ and sensitivity $k$ using the gravitational acceleration as reference quantity. The sensor output is recorded while the robot is put in five different positions, once on each side and upright. Then the mean value is calculated for each of the five sequences. To account for possible misalignment of the sensor chip and possible crosstalk of the two perpendicular measurement channels on each chip, also a cross-sensitivity is modeled.

A simple linear sensor model is given by

$$y_i = b_i + k_{i1}x_1 + k_{i2}x_2, \tag{4.1}$$

where $y$ denotes the sensor output and $x$ denotes the true quantity to be measured. $i \in [1; 2]$ denotes the considered channel.

Collecting the output of the two channels for the five reference measurements in the vectors $\boldsymbol{y}_1, \boldsymbol{y}_2 \in \mathrm{R}^{5\times 1}$ yields the over-determined system of equations with full rank three

$$\boldsymbol{y}_i = \underbrace{\begin{bmatrix} 1 & 9.81 & 0 \\ 1 & 0 & 9.81 \\ 1 & -9.81 & 0 \\ 1 & 0 & -9.81 \\ 1 & 0 & 0 \end{bmatrix}}_{\boldsymbol{X}} \underbrace{\begin{bmatrix} b_i \\ k_{i1} \\ k_{i2} \end{bmatrix}}_{\boldsymbol{p}_i} \qquad (4.2)$$

whose least-squares solution is given by

$$\boldsymbol{p}_i = (\boldsymbol{X}^\mathrm{T}\boldsymbol{X})^{-1}\boldsymbol{X}^\mathrm{T}\boldsymbol{y}_i. \qquad (4.3)$$

To validate the model assumptions, the magnitude of the residuals $\boldsymbol{e}$ is checked,

$$\boldsymbol{e}_i = \boldsymbol{y}_i - \boldsymbol{X}\boldsymbol{p}_i. \qquad (4.4)$$

Finally, the true quantities $x_i$ are given by

$$\begin{bmatrix} x_1 \\ x_2 \end{bmatrix} = \begin{bmatrix} k_{11} & k_{12} \\ k_{21} & k_{22} \end{bmatrix}^{-1} \begin{bmatrix} y_1 - b_1 \\ y_2 - b_2 \end{bmatrix}, \qquad (4.5)$$

as linear functions of the measurement outputs $y_i$. Full rank of the system matrix is practically guaranteed since the channels are orthogonal and the crosstalk is within reasonable bounds.

### 4.3.2 Calibration of the gyro-sensor

Again, a linear characteristic is assumed, the bias is determined as the mean value of the recorded sensor signal while the robot is not moving. The sensitivity is found by recording the sensor signal while rotating the robot on a laboratory servo device with adjustable angular velocity. Special attention has to be directed towards the alignment of the robot during this calibration procedure: a misalignment of the yaw-axis and the rotation axis of the servo-device would falsify the sensitivity.

### 4.3.3 Calibration of the encoders

The encoders actually measure the rotation angle of the wheels. Velocity information is only obtained by numerical differentiation, i.e. by counting the number of steps during one sampling interval. Therefore, they are definitely bias-free. Their sensitivity $k$ in counted steps per rads$^{-1}$ is determined by the number of steps $N_{\mathrm{Enc}}$ during one motor revolution, i.e. the encoder resolution, the transmission ratio $n$ between motor and wheel and the sampling time $T_\mathrm{N}$, and is given by

$$k = \frac{N_{\mathrm{Enc}}}{2\pi} n T_\mathrm{N}. \qquad (4.6)$$

## 4.4 Introduction to the developed algorithm

Inspired by Borenstein and Feng's paper about gyrodometry, [15], the general concept of the developed navigation algorithm is to substitute data from inertial sensors for odometric data only during periods where odometry is likely to be unreliable, i.e. when tangential slip or side-slip occur.

The primary task is therefore to detect these situations. As functions of appropriate criteria, mixing ratios are calculated, by which a combination ranging from purely inertial estimation to purely odometric estimation is defined. Thereby the advantages of both types of sensors are exploited: Inertial measurement is not affected by wheel slippage, odometric measurement is not affected by bias, when properly calibrated.

The secondary task is the quantitatively reliable estimation of the track speed, the yaw rate, the side-slip angle and the posture consisting of the cartesian coordinates and the orientation relative to some inertial coordinate system to be fed to the predictive control algorithm.

As a by-product of slip detection, slip control becomes possible. Whenever excessive tangential slip or side-slip is detected, the pre-planned trajectory is transiently altered to reduce track speed and accelerations, much in the way as commonly found in modern automotive applications under notions such as electronic stability program, anti-lock brake or traction slip control.

The most important consequence of slip control is that it also helps increase the accuracy of navigation, because it narrows the periods where inertial navigation must be substituted for odometric navigation.

The algorithm presented in this chapter is also documented in a conference contribution to the 2006 IEEE International Conference on Robotics, Automation and Mechatronics [83] and in a journal article in Robotics and Autonomous Systems [52].

## 4.5 State estimation

### 4.5.1 Inertial navigation

Inertial navigation processes the data from the five inertial sensor signals at $T_N = 0.002$s, the sampling time of the navigation algorithm. The corresponding sampling instants are indicated by $k$. The sampling time of the control algorithm is $T_C = 0.02$s and the corresponding sampling instants are indicated by $K$.

The redundancy of the two respective acceleration measurements is used to reduce the statistical variance of its noise by calculating their arithmetic mean, thereby effectively yielding three measurements instead of five.

The statistical variance of a zero-mean signal is defined as the expectation of its square,

$$\sigma^2 = \mathrm{E}\{x^2\}, \tag{4.7}$$

where E{·} denotes the expectation operator. Under the assumption that the noise variances of the two signals are equal, i.e. $\sigma_1^2 = \sigma_2^2$ and that the noise of the two signals is uncorrelated, the noise variance of the arithmetic mean of the two signals is shown to be half of the noise variance of the original signals,

$$\sigma^2 = E\left\{\left(\frac{x_1 + x_2}{2}\right)^2\right\} = \frac{1}{4}E\{(x_1^2 + 2x_1x_2 + x_2^2)\} = \frac{1}{4}(E\{x_1^2\} + \underbrace{E\{2x_1x_2\}}_{0} + E\{x_2^2\}) = \frac{1}{4}(\sigma_1^2 + \sigma_2^2) = \frac{\sigma_1^2}{2}, \quad (4.8)$$

where the linearity property of the expectation operator is used.
The robot's felt shoes are not entirely free of play. Therefore, the acceleration sensors record a certain additional acceleration due to gravity whenever the robot is tilted forwards or backwards. Compensation of this static acceleration contribution is accomplished by

$$a_{t,\text{comp}} = \text{sign}a_t \max(|a_t| - g \sin \varepsilon, 0), \quad (4.9)$$

where $\varepsilon$ is a tilt angle of in practice at most a few degrees.
In a preliminary version of the navigation algorithm, Extended Kalman filtering was employed for inertial navigation. A detailed derivation of the EKF is found in Appendix B.
To be able to incorporate acceleration information, a dynamic model including the absolute accelerations of the robot is needed. A continuous-time dynamic model including all measurements but with the smallest possible number of states is given by

$$\begin{bmatrix} \dot{v}_t \\ \dot{v}_n \\ \dot{\omega} \\ \dot{a}_t \\ \dot{a}_n \end{bmatrix} = \begin{bmatrix} a_t + v_n\omega \\ a_n - v_t\omega \\ 0 \\ 0 \\ 0 \end{bmatrix}, \quad (4.10)$$

where $v_t$, $v_n$, $\omega$ and the absolute accelerations $a_t$ and $a_n$ are chosen as the states. The system's inputs are not modelled.
Discretisation using a simple forward-Euler-scheme yields the discrete state prediction

$$\hat{\boldsymbol{x}}_{k|k-1} = \begin{bmatrix} \hat{a}_{t,k-1} + \hat{v}_{n,k-1}\hat{\omega}_{k-1} \\ \hat{a}_{n,k-1} - \hat{v}_{t,k-1}\hat{\omega}_{k-1} \\ 0 \\ 0 \\ 0 \end{bmatrix} T_N + \underbrace{\begin{bmatrix} \hat{v}_{t,k-1} \\ \hat{v}_{n,k-1} \\ \hat{\omega}_{k-1} \\ \hat{a}_{t,k-1} \\ \hat{a}_{n,k-1} \end{bmatrix}}_{\hat{\boldsymbol{x}}_{k-1}}. \quad (4.11)$$

The index $k|k-1$ can be spelled out as *prediction of the states at instant $k$, based on data of instant $k-1$*.
The measurement vector is

$$\boldsymbol{y}_k = \begin{bmatrix} \omega_k \\ (a_{t1,k} + a_{t2,k})/2 \\ (a_{n1,k} + a_{n2,k})/2 \end{bmatrix}. \quad (4.12)$$

## 4.5. STATE ESTIMATION

The prediction of the measurements at instant $k$ from the state estimate at instant $k-1$ is given as

$$\hat{\boldsymbol{y}}_{k|k-1} = \begin{bmatrix} \hat{\omega}_{k|k-1} \\ \hat{a}_{\text{t},k|k-1} \\ \hat{a}_{\text{n},k|k-1} \end{bmatrix}. \tag{4.13}$$

The Jacobian of the nonlinear discrete state space representation (4.11) is given by

$$\boldsymbol{A}_{k-1} = \begin{bmatrix} 1 & \hat{\omega}_{k-1}T_\text{N} & \hat{v}_{\text{n},k-1}T_\text{N} & T_\text{N} & 0 \\ -\hat{\omega}_{k-1}T_\text{N} & 1 & -\hat{v}_{\text{t},k-1}T_\text{N} & 0 & T_\text{N} \\ 0 & 0 & 1 & 0 & 0 \\ 0 & 0 & 0 & 1 & 0 \\ 0 & 0 & 0 & 0 & 1 \end{bmatrix}. \tag{4.14}$$

The measurement prediction (4.13) is linear in the states and its Jacobian with respect to the states is therefore given by the constant matrix

$$\boldsymbol{C} = \begin{bmatrix} 0 & 0 & 1 & 0 & 0 \\ 0 & 0 & 0 & 1 & 0 \\ 0 & 0 & 0 & 0 & 1 \end{bmatrix}. \tag{4.15}$$

The covariance of the prediction $\hat{\boldsymbol{x}}_{k|k-1}$ is given by

$$\boldsymbol{P}_{k|k-1} = \boldsymbol{A}_{k-1}\boldsymbol{P}_{k-1}\boldsymbol{A}_{k-1}^\text{T} + \boldsymbol{Q}, \tag{4.16}$$

where $\boldsymbol{Q} \in \mathbb{R}^{5\times 5}$ denotes the constant diagonal covariance matrix of the state uncertainty, whose entries are tuned to be as small as possible but large enough to ensure convergence.
The Kalman gain matrix $\boldsymbol{K}_k \in \mathbb{R}^{5\times 3}$ is obtained by minimisation of the trace of the state covariance matrix $\boldsymbol{P}_k \in \mathbb{R}^{5\times 5}$,

$$\boldsymbol{K}_k = \boldsymbol{P}_{k|k-1}\boldsymbol{C}^\text{T}(\boldsymbol{C}\boldsymbol{P}_{k|k-1}\boldsymbol{C}^\text{T} + \boldsymbol{R})^{-1}, \tag{4.17}$$

where $\boldsymbol{R} \in \mathbb{R}^{3\times 3}$ is the constant covariance of the sensor noise.
Using the Kalman gain matrix $\boldsymbol{K}_k$ and the difference between predicted and actual measurement, the states are updated to obtain the final state estimate,

$$\boldsymbol{x}_k = \boldsymbol{x}_{k|k-1} + \boldsymbol{K}_k(\boldsymbol{y}_k - \hat{\boldsymbol{y}}_{k|k-1}). \tag{4.18}$$

As a last step, the covariance of the updated state estimate

$$\boldsymbol{P}_k = (\boldsymbol{I} - \boldsymbol{K}_k\boldsymbol{C})\boldsymbol{P}_{k|k-1}(\boldsymbol{I} - \boldsymbol{K}_k\boldsymbol{C})^\text{T} + \boldsymbol{K}_k\boldsymbol{R}\boldsymbol{K}_k^\text{T} \tag{4.19}$$

is calculated to be used in the next sampling instant.

This concept works well, a closer investigation, however, shows that the statistical variance of the estimate of the velocity components grows roughly linearly. This behaviour corresponds to a discrete integration where the integrand is affected by noise, according to a Gauss-Markov-sequence.
Moreover, the entries of the Kalman gain matrix corresponding to the velocity components remain very close to zero at all times. Therefore, virtually no update takes place. The only

information about the velocity components stems from the prediction step. Again, this corresponds to a pure integration.
By evaluating the $mn \times n$-observability matrix, whose rank must be equal to the number of states $n = 5$,

$$\mathcal{O}_k = \begin{bmatrix} C \\ CA_k \\ \vdots \\ CA_k^{n-1} \end{bmatrix} \qquad (4.20)$$

a double rank deficiency is observed, which is independent of the state. This is entailed by the structure of the system Jacobian $A_k$ and the measurement matrix $C$. This circumstance is common to all systems containing states which can only be obtained by integration.
Therefore, the application of the EKF does not provide any advantages compared to a simple discrete integration, given by

$$\begin{aligned}
\hat{v}_{tk} &= \hat{v}_{tk-1} + T_N\Big((a_{t1k-1} + a_{t2k-1})/2 + \hat{v}_{nk-1}\omega_{k-1}\Big) \\
\hat{v}_{nk} &= \hat{v}_{nk-1} + T_N\Big((a_{n1k-1} + a_{n2k-1})/2 - \hat{v}_{tk-1}\omega_{k-1}\Big) \\
\hat{\omega}_k &= \omega_k.
\end{aligned} \qquad (4.21)$$

With regard to the higher complexity and increased calculation time, the EKF can therefore be replaced by discrete integration.

### 4.5.2 Odometry

To obtain tangential velocity $v_t$ and yaw rate $\omega$ from the measured wheel angular velocities, a simple geometric relation is applied,

$$\begin{bmatrix} v_{t,k} \\ \omega_k \end{bmatrix} = \begin{bmatrix} r/2 & r/2 \\ r/b & -r/b \end{bmatrix} \begin{bmatrix} \omega_{R,k} \\ \omega_{L,k} \end{bmatrix} \qquad (4.22)$$

The parameters wheel radius $r$ and wheelbase $b$ have to be calibrated, especially the latter, because the distance between the wheels' contact points cannot clearly be defined.
Calibration is performed using Borenstein and Feng's method [16], where a square-shaped path is executed both clockwise and counterclockwise a number of times. For both, the centroids of the end-points are calculated, which should of course be identical with the starting point. From the positions of the centroids relative to the starting point, the ratio between the wheels' radii on the one hand and the wheelbase on the other hand can be determined: Miscalibration of the ratio between the wheels' radii causes a curved path instead of a straight line leading to an asymmetric error with respect to the direction of rotation, whereas miscalibration of the wheelbase causes over- or understeering in corners leading to a symmetric error.
The mean value of the wheels' radii is determined by executing a straight path and measuring the distance between starting point and end point.
As long as no non-systematic errors occur, odometric measurement is very accurate over long distances by means of the described calibration procedure.
Naturally, from odometric measurement no information about the side-slip velocity $v_n$ can be obtained.

## 4.5. STATE ESTIMATION

### 4.5.3 Tangential slip detection

A batch of $T_C/T_N = 10$ samples of sensor data is processed. The average tangential velocity difference of the current batch after an initial reset is calculated by discrete integration of the arithmetic mean of the tangential accelerations (upper left index O stands for odometric estimate)

$$\overline{\Delta v}_K = \frac{T_N}{T_C} \sum_{k=1}^{T_C/T_N} \left( {}^O v_{t,k} - {}^O v_{t,K-1} - \sum_{j=1}^{k}(a_{t1,j} + a_{t2,j})\frac{T_N}{2} \right), \qquad (4.23)$$

which effectively means that a trend is detected, but not a stationary slip. In this approximation, the Coriolis-term is simply omitted. Thereby the criterion can be explicitly calculated instead of having to be calculated recursively, which saves calculation time.

### 4.5.4 Mixing

The final estimate of the states is generated by linear mixing between odometric (upper left index O) and inertial (upper left index I) estimation.

As the criterion for mixing of the tangential velocity $v_t$ the tangential slip is used. The mixing parameter is calculated every $K$-th instant by

$$C_{1,K} = p\, C_{1,K-1} + (1-p)\underbrace{\frac{1}{1 + \exp(k_1(|\overline{\Delta v}_K| - b_1))}}_{\tilde{C}_{1,K}}, \qquad (4.24)$$

a double sided logsig-function as depicted in Fig. 4.2. Additionally, $C_1$ is dynamically relaxed following discrete first order delay behavior. This measure proved necessary during practical testing. It ensures a greater level of continuity of $C_1$ over time, which has a stabilising effect on the overall performance.

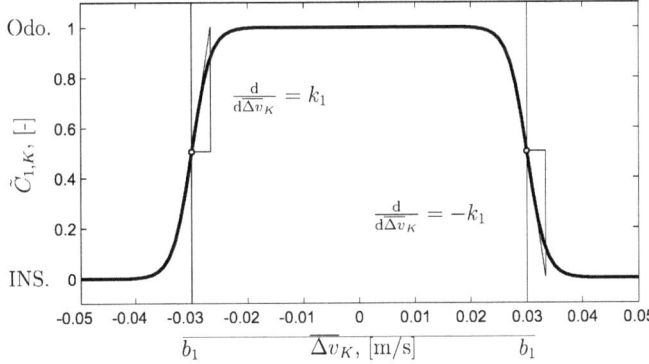

Figure 4.2: Double-sided logsig-function as switch between odometric (Odo.) and inertial (INS.) navigation.

The tuning parameters are obtained as follows:

- $k_1$: Steepness, chosen as steep as possible, but as flat as necessary to avoid limit cycle behavior.

- $b_1$: Threshold, chosen as small as possible, but large enough to exclude spurious slip detection due to acceleration sensor bias.

- $p$: Discrete first order delay pole, chosen as close to zero, i.e. as fast as possible, but large enough to avoid limit cycle behavior.

For the lateral velocity $v_\mathrm{n}$, the absolute lateral acceleration is used as a criterion, its mixing parameter computed every $k$-th instant by

$$C_{2,k} = \frac{1}{1 + \exp(k_2(|\hat{a}_{\mathrm{n},k|k}| - b_2))}, \tag{4.25}$$

where the parameters $k_2$ and $b_2$ are obtained much in the same way as for the tangential velocity. The criterion for yaw-rate mixing is simply the absolute difference between odometric and inertial measurement, [15]. If the absolute difference becomes exceedingly large, odometric measurement is assumed to be unreliable, therefore the mixing parameter is obtained every $k$-th instant by

$$C_{3,k} = \frac{1}{1 + \exp(k_3(|^\mathrm{O}\omega_k - {}^\mathrm{I}\omega_k| - b_3))} \tag{4.26}$$

with parameters $k_3$ and $b_3$ as above.
The linear mixing laws for tangential velocity $v_\mathrm{t}$,

$$\hat{v}_{\mathrm{t},k} = C_{1,K}{}^\mathrm{O}v_{\mathrm{t},k} + (1 - C_{1,K})^\mathrm{I}v_{\mathrm{t},k} \tag{4.27}$$

and yaw rate $\omega$,

$$\hat{\omega}_k = C_{3,k}{}^\mathrm{O}\omega_k + (1 - C_{3,k})^\mathrm{I}\omega_k, \tag{4.28}$$

are straightforward. In contrast, the lateral velocity $v_\mathrm{n}$ is calculated by

$$\hat{v}_{\mathrm{n},k} = C_{2,k}\hat{v}_{\mathrm{n},k-1}q + (1 - C_{2,k})^\mathrm{I}v_{\mathrm{n},k}, \tag{4.29}$$

so that it is taken fully from inertial navigation when the absolute lateral acceleration is beyond the threshold $b_2$ but geometrically relaxed towards zero when the lateral acceleration is within the threshold. This is necessary because no information about the lateral velocity can be obtained from odometry, i.e. ${}^\mathrm{O}v_{\mathrm{n},k}$ does not exist. Without relaxation, however, frequent crossing of the threshold due to inevitably noisy lateral acceleration would result in a frequent reset of the lateral velocity to zero.

The factor $q \in [0; 1]$ of geometric decay is chosen as small as possible, i.e. as fast as possible, but large enough to ensure consistent velocity measurement!

## 4.5.5 Calculation of the side-slip angle

Every $K$ the tangential and normal velocity are transformed into the equivalent description by track speed and side-slip angle, which serve as the input to the predictive control algorithm described in Chapter 3.

The signed track speed is given by

$$\hat{v}_K = \sqrt{\hat{v}_{t,K}^2 + \hat{v}_{n,K}^2}\,\text{sign}(\hat{v}_{t,K}), \tag{4.30}$$

and the side-slip angle, defined in the interval $[-\frac{\pi}{2}; \frac{\pi}{2}]$ is calculated from the velocity components using the inverse tangent,

$$\alpha_K = \text{atan}\left(\frac{\hat{v}_{n,K}}{\hat{v}_{t,K}}\right) f(\hat{v}_K, \hat{\omega}_K), \tag{4.31}$$

where the same validity function is used as in Section 2.4.1.

## 4.5.6 Integration of the posture

Finally, the positions and orientations are calculated by integration using a simple-forward-Euler-discretised representation of the kinematics (2.1),

$$\begin{aligned} x_{k+1} &= x_k + (v_{t,k}\cos\varphi_k - v_{n,k}\sin\varphi_k)T_N \\ y_{k+1} &= y_k + (v_{t,k}\sin\varphi_k + v_{n,k}\cos\varphi_k)T_N \\ \varphi_{k+1} &= \varphi_k + \omega_k T_N. \end{aligned} \tag{4.32}$$

# 4.6 Slip control

The reference input to the predictive control algorithm executed at a sampling frequency of $1/T_C$ consists of future reference positions and orientations up to a prediction horizon, which are generated by a trajectory planning algorithm.

In case of excessive slip, slip control can now transiently override the trajectory given by the planning algorithm by interpolating the original trajectory points with a reduced step size, thus remaining on the same geometric path, but with a certain delay and reduced velocity and accelerations. When no tangential or lateral slip occurs, the step size is increased beyond one to re-match the original trajectory.

The step size $\delta_K$, also based on the previous value $\delta_{K-1}$ to ensure continuity, is computed every $K$-th instant by

$$\delta_K = \min(\max(\delta_{K-1} \cdot \left(\delta_{\min} + \min\left(C_{1,K}, \frac{1}{1+\exp(k_4\,(|\alpha_K| - b_4))}\right)\cdot\left(q - \delta_{\min}\right)\right), \delta_{\min}), \delta_{\max,K}), \tag{4.33}$$

where the maximum step size $\delta_{\max,K}$ is given by

$$\delta_{\max,K} = \begin{cases} \delta_{\max}(>1) & : & K - \Pi_K > c \\ 1 & : & c > K - \Pi_K > d \\ \delta_{\text{rcm}}(<1) & : & d > K - \Pi_K. \end{cases} \tag{4.34}$$

The complicated structure enables instantaneous reaction to excessive tangential slip or side-slip, determined by the expression $\min(C_{1,K}, \frac{1}{1+\exp(k_4(|\alpha_K|-b_4))})$, which ranges from zero to one. If this expression assumes a value of one, i.e. neither tangential nor side-slip occur, the step size grows geometrically according to $\delta_K = \delta_{K-1}(\delta_{\min} + 1(q - \delta_{\min})) = \delta_{K-1}q$. If a value of zero is assumed, the step size drops according to $\delta_K = \delta_{K-1}\delta_{\min}$. The step size is limited to the interval $[\delta_{\min}, \delta_{\max,K}]$. Thereby the step size drops instantaneously to $\delta_{\min}$ if required, whereas its growth can be tuned to be much slower, thus avoiding excessive acceleration which could in turn lead to repeated occurance of slip.
The position on the preplanned trajectory is then given as

$$\Pi_{K+1} = \Pi_K + \delta_K, \tag{4.35}$$

whereas the nominal integer position on the preplanned trajectory naturally equals $K$.
A number of parameters determine the behavior:

- $\delta_{\min} < 1$: Minimum step size, determines the reduction of velocity in case of slip.
- $\delta_{\max} > 1$: Maximum step size, determines how quickly the original trajectory is recovered.
- $q$: Factor of geometric growth of the step size after reduction, if it is chosen too large, limit cycle behavior can occur.
- $k_4, b_4$: Steepness and threshold for the side-slip angle criterion.
- $\delta_{\text{rcm}} < 1$: Step size to match reference trajectory after overshoot.
- $c, d$: Upper and lower threshold for re-matching.

With the notation ${}^T\underline{x}_{\text{ref},i} := {}^T x_{\text{ref}}(\text{floor}(\Pi_K + i\delta_K))$ and ${}^T\overline{x}_{\text{ref},i} := {}^T x_{\text{ref}}(\text{ceil}(\Pi_K + i\delta_K))$ the reference positions and orientations finally fed to the predictive control algorithm are obtained by interpolation according to

$${}^C x_{\text{ref},K+i} = {}^T\underline{x}_{\text{ref},i} + (\Pi_K + i\delta_K - \text{floor}(\Pi_K + i\delta_K))({}^T\overline{x}_{\text{ref},i} - {}^T\underline{x}_{\text{ref},i}), \tag{4.36}$$

where the upper left index T denotes original data from trajectory planning and C denotes data finally fed to the control algorithm.
Thus the history of the velocities associated with the reference trajectory is expanded or compressed according to $\delta_K$ with equal area, which means that both absolute velocity and absolute accelerations are reduced. In Fig. 4.3 the principle is illustrated for a straight line.

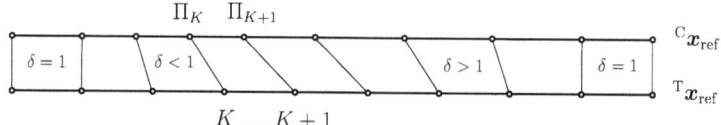

Figure 4.3: Principle of trajectory interpolation.

## 4.7 Block diagram

In Fig. 4.4 a complete block diagram of the navigation and slip control algorithm is depicted.

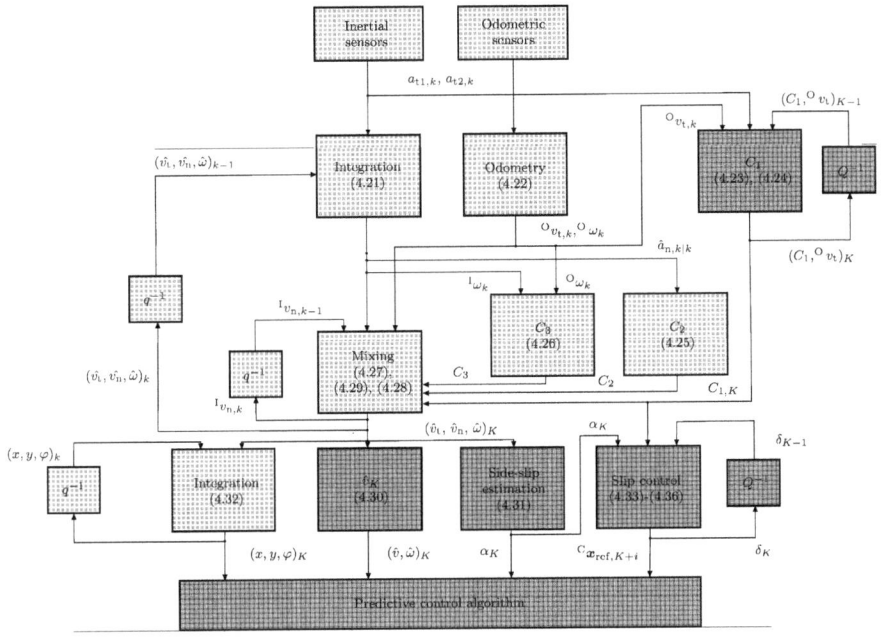

Figure 4.4: Block diagram, light grey: $T_\mathrm{N}$, dark grey: $T_\mathrm{C}$.

# Chapter 5

# Motion planning

## 5.1 Introduction

Mobile robot motion planning is usually formulated as a hierachical sequence of path planning and trajectory generation, as stated in a comprehensive volume by Latombe [61]. First, a sequence of waypoints is generated, which are subsequently linked by trajectory or path sections. In the latter case, a velocity profile must be superimposed to obtain a trajectory suitable to be fed to a control algorithm.

For motion planning, some internal representation of the environment is required. For autonomous operation, however, complete a-priori knowledge of the environment cannot be premised. Therefore, it is necessary to construct a map of the robot's surroundings within its memory, based on the perception of the environment. This procedure is called map-building. Among the possible sensing mechanisms are laser scanning, ultrasonic range measurement and radar, providing so-called range images. Another possibility for environment perception is stereo vision.

A completely autonomous mobile robot must therefore be capable of map-building, waypoint generation and trajectory generation.

## 5.2 State of the art

### 5.2.1 Waypoint generation

A vast number of publications address waypoint generation. In principle they can be assigned to the major strands roadmap methods, cell decomposition methods and potential field methods. Usually complete knowledge of the environment is required, [61].

Roadmap methods use the vertices of a polygonal representation of the environment to construct a graph. From this graph, the shortest feasible connection to the goal configuration is sought. Wang et al. e.g. [98] use a genetic algorithm to find a path. Aydin and Temeltas [7] use the well-known visibility graph method.

Cell decomposition methods partition the environment into polygons that either belong to the free space or to obstacles. The centroids of neighbouring free segments can then be used as a sequence of waypoints. Examples for approximate cell decomposition, often called quadtree decomposition, are reported by Burlet et al. in [20] and Jacobs and Canny in [51].

Potential field methods construct an artificial potential field around the goal, where the goal constitutes the global minimum. Obstacles are represented by increasing the value of the potential field. By evaluating the gradient of the potential field, a sequence of waypoints towards the goal is found. Hussein and Elnagar exploit the properties of Maxwell's equations in [48] for three-dimensional potential field planning, Barraquand et al. [9] use numerically constructed potential fields. A major problem with potential field methods are local minima generated by unfavourable obstacle configurations.

Other authors provide results for partially known environments. Elnagar and Basu [29] present a heuristic approach based on the simultaneous maximisation of a safety function incorporating local environment information and goal attraction. Ersson and Hu [31] use a grid-based representation of the local environment to determine a waypoint among a number of discrete candidates. Ferguson and Stentz [35, 87] present a well-known algorithm called D*, where an initially generated path is modified during motion whenever new environment information is acquired. Therefore, this approach is called path replanning. This algorithm also uses a grid-based environment representation.

### 5.2.2 Path planning

Many authors employing explicit geometric descriptions for their paths rely on the fundamental results by Dubins [27] and much later by Reeds and Shepp [79], where terminal configurations are linked by sequences of circular arcs and straight lines, in the latter case including reversals. Among those researchers are Mirtich and Canny [68], introducing the notion *shortest feasible path*, where the nonholonomic constraints of a wheeled mobile robot are captured in a modified metric. Fraichard, Scheuer and Laugier in [40] and other work modify the original proposal by Reeds and Shepp to obtain curves with continuous curvature. Laumond et al. use Reeds and Shepp-curves in [62] to update a previously generated path such that the nonholonomic constraint is respected.

Piazzi, Guarino lo Bianco et al. in [78] and other work and Villagra and Mounier in [95] use quintic splines to interpolate the trajectory described by cartesian coordinates between initial and terminal configuration, while optimising the smoothness of the curve, i.e. minimising the curvature derivative.

Kanayama and Hartman [56] also employ the curvature and the curvature derivative as optimisation criteria, but their waypoints are connected by cubic spirals.

### 5.2.3 Trajectory planning

So far, only path planning has been discussed. Fewer authors address the problem of dynamic trajectory generation.

Fraichard uses a discretised *state-time space* in [38] to obtain a dynamic trajectory. Fraichard and Laugier [39] use an approach called path-velocity decomposition to first plan a path and then overlay it with a velocity profile.
Haddad *et al.* [43] present an approach where so-called milestone nodes, a synonym to waypoints, are connected via splines. The trajectory is represented in cartesian coordinates. The geometry and the velocity profile is iteratively optimised, but with two distinct optimisation steps.
Hussein and Elnagar [47] use analytical calculus of variations to optimise a trajectory represented by parameterised arc length and orientation angle. In [30] they present an extension to three-dimensional environments.

### 5.2.4 Map-building

A comprehensive survey by Thrun [91] gives an overview over different existing approaches to map-building. Virtually all robotic mapping algorithms are probabilistic, some employ Kalman filtering to incorporate newly acquired data into the map.
A great amount of research effort is being directed towards Simultaneous Localisation and Map-building (SLAM), where the features in the map are used for exteroceptive navigation and the map is advanced at the same time. Guivant and Nebot [42] and Kelly and Unnikrishnan [58] present different approaches to SLAM.
Thrun *et al.* [92] present a probabilistic map-building algorithm based on a detailed description of the uncertainty of odometric navigation.
Leonard *et al.* [64] employ Extended Kalman filtering to map-building, where in addition to the noise covariance of a measurement a validation procedure is developed.
Borges and Aldon [17] present an alternative approach to SLAM without EKF. They also give some details and references to feature extraction. From a laser scan, a set of line segments is extracted, supported by feature extraction from an optical image.

## 5.3 Introduction to the developed algorithms

The developed motion planning concept consists of a fully dynamic trajectory generation algorithm, a waypoint generator which works with incomplete environment information and a simplified map-building scheme. Contrary to most existing approaches, the presented trajectory generator is capable of simultaneous optimisation of geometric and dynamic properties. This is accomplished by multi-objective numerical optimisation.

The first crucial novelty is the description of the trajectory by tangential velocity and curvature as functions of time. The second innovation is to linearly superpose basis functions for these descriptive quantities. Thereby the problem is transferred from analytical calculus of variations to multi-objective nonlinear constrained optimisation, thus limiting the otherwise infinite function space to the set of basis functions.

The choice of tangential velocity and curvature as descriptive quantities has proven to be most feasible, and is easily motivated by comparison with the behaviour of a human driver, who controls the motion of a vehicle basically by accelerator and brake, corresponding to the tangential velocity, and the steering wheel, corresponding to the curvature.

A geometric criterion for optimisation is the total path length, the dynamic criteria are the lateral acceleration, the tangential acceleration derivative, the curvature derivative corresponding to the steering wheel's angular velocity and therefore also called steering rate, and the time to reach the target.

While those criteria are optimised according to some previously defined trade-off, a number of hard constraints are exactly met. Naturally, the desired position and orientation of the end-point have to be met. Velocity, acceleration and curvature at the starting point of the trajectory are prescribed to ensure continuity, velocity and acceleration at the endpoint are prescribed to enable jerk-free stopping. Moreover, for any configuration between the starting point and the end point of the trajectory, the minimal obstacle clearance is met.

The presented trajectory generation algorithm is generically applicable to any nonholonomic vehicle, this includes the classical car-like vehicle and is not limited to the two-wheeled differentially driven vehicle so far addressed in this work.

The obstacles are represented by polygons, it is assumed that this type of representation can be derived e.g. from a laser range image. This assumption is supported by the results of Borges and Aldon [17]. At each sampling instant the currently visible polygons are merged with previously known environment information, thereby building a map of the environment whilst moving within it. This corresponds to a simplified deterministic map-building scheme under disregard of localisation uncertainty and possibly noisy or spurious sensor data. This serves for the validation of the waypoint generator and the trajectory planner, but is not by itself a focus of research of this work.

The waypoint generator is closely connected to the chosen environment representation. It relies primarily on the currently visible obstacles, but also incorporates already available environment information. Unlike the majority of documented path planning approaches this concept does not require full knowledge about the environment, but on the other hand does not necessarily provide geometric optimality. It is believed that this approach is best suited to the practical requirements of fully autonomous operation, where changing or entirely unknown environments are inevitable.

The algorithms presented in this chapter are also documented in a conference contribution to the 2007 American Control Conference, [84].

## 5.4 Trajectory generation

### 5.4.1 Basis functions and descriptive quantities

The basis functions are defined as functions of dimensionless time $\tau$ on the interval $\tau \in [0,1]$, evaluated at $m$ evenly spaced nodes. As basis function prototype the logsig-function is chosen,

$$\mathrm{logsig}(k\tau + d) := \frac{1}{1 + \exp(-k\tau - d)}, \tag{5.1}$$

whose derivative is given by

$$\frac{\mathrm{d}}{\mathrm{d}\tau}\mathrm{logsig}(k\tau + d) = k\,\mathrm{logsig}(k\tau + d)\Big(1 - \mathrm{logsig}(k\tau + d)\Big). \tag{5.2}$$

where $k$ is the steepness and $d$ is a time-shift. The set of basis functions is composed of a number of evenly spaced prototypes with equal steepness, as well as possibly a linear and a constant function.

The basic descriptive quantities, given by $m$-vectors (curvature, steering rate, velocity, acceleration and acceleration change) are computed straightforward by linear superposition according to the parameters contained in the $n_{v,\kappa}$-parameter-vectors $\boldsymbol{p}_v$ and $\boldsymbol{p}_\kappa$,

$$\begin{aligned} \boldsymbol{\kappa} &= \boldsymbol{F}_\kappa^\mathrm{T} \boldsymbol{p}_\kappa & \dot{\boldsymbol{\kappa}} &= \boldsymbol{F}_\kappa'^\mathrm{T} \boldsymbol{p}_\kappa \frac{1}{T} \\ \boldsymbol{v} &= \boldsymbol{F}_v^\mathrm{T} \boldsymbol{p}_v \frac{1}{T} & \boldsymbol{a}_\mathrm{t} &= \boldsymbol{F}_v'^\mathrm{T} \boldsymbol{p}_v \frac{1}{T^2} & \dot{\boldsymbol{a}}_\mathrm{t} &= \boldsymbol{F}_v''^\mathrm{T} \boldsymbol{p}_v \frac{1}{T^3}, \end{aligned} \tag{5.3}$$

introducing the time dimension by dividing by the entire period $T$ of the trajectory in the correct power. The matrices $\boldsymbol{F}_{v,\kappa} \in \mathbb{R}^{n_{v,\kappa} \times m}$ and $\boldsymbol{F}'_{v,\kappa} \in \mathbb{R}^{n_{v,\kappa} \times m}$, which are computed off-line, contain the basis functions and their derivatives respectively, numerically evaluated at the $m$ nodes.

Yaw rate and lateral acceleration are composite quantities, given by

$$\boldsymbol{\omega} = \boldsymbol{\kappa} \circ \boldsymbol{v}, \quad \boldsymbol{a}_\mathrm{n} = \boldsymbol{\omega} \circ \boldsymbol{v} = \boldsymbol{\kappa} \circ \boldsymbol{v} \circ \boldsymbol{v}, \tag{5.4}$$

where the symbol $\circ$ denotes the Schur-Hadamard vector product (elementwise multiplication). The orientation angle (heading) $\varphi$ is calculated by approximate discrete integration over $t \in [0;T]$, where the discrete integration step width is given by $\Delta t = T\Delta\tau$ and $\Delta\tau = 1/(m-1)$, according to

$$\boldsymbol{\varphi}^\mathrm{T} = \left[ 0,\ \omega_1 T\Delta\tau,\ \ldots,\ \sum_{k=1}^{m-1} \omega_k T\Delta\tau \right]. \tag{5.5}$$

The cartesian coordinates $x$ and $y$ are calculated by further integrating the well-known nonholonomic kinematics of the rolling wheel,

$$\begin{aligned} \boldsymbol{x}^\mathrm{T} &= \left[ 0,\ v_1 \cos\varphi_1 T\Delta\tau,\ \ldots,\ \sum_{k=1}^{m-1} v_k \cos\varphi_k T\Delta\tau \right], \\ \boldsymbol{y}^\mathrm{T} &= \left[ 0,\ v_1 \sin\varphi_1 T\Delta\tau,\ \ldots,\ \sum_{k=1}^{m-1} v_k \sin\varphi_k T\Delta\tau \right]. \end{aligned} \tag{5.6}$$

## 5.4.2 Performance functional

For multi-objective optimisation, a positive definite scalar performance functional $J$ is constructed, containing the integral mean of the squared lateral acceleration, the squared steering rate and the squared acceleration rate, the squared total time and the absolute path length, weighted by coefficients $\alpha_1$ through $\alpha_5$, thus effectively combining dynamic and geometric criteria.

The aim is to link the initial and terminal configurations by trajectories as short (absolute path length) and smooth (steering rate) as possible in the shortest possible time, while keeping the (mean) lateral acceleration as low as possible. The $n = n_v + n_\kappa + 1$ parameters to be optimised are $\boldsymbol{p}^\mathrm{T} = [\boldsymbol{p}_v^\mathrm{T}, \boldsymbol{p}_\kappa^\mathrm{T}, T]$. $J$ is analytically given by

$$J = \alpha_1 \frac{1}{T} \int_0^T a_\mathrm{n}(t)^2 \mathrm{d}t + \alpha_2 \frac{1}{T} \int_0^T \dot\kappa(t)^2 \mathrm{d}t + \alpha_3 \frac{1}{T} \int_0^T \dot{a}_\mathrm{t}(t)^2 \mathrm{d}t + \alpha_4 \int_0^T |v(t)| \mathrm{d}t + \alpha_5 T^2. \quad (5.7)$$

Its discrete approximation in terms of the descriptive quantities (5.3) and (5.4) then reads

$$J \approx \alpha_1 \underbrace{\boldsymbol{a}_\mathrm{n}^\mathrm{T} \boldsymbol{a}_\mathrm{n} \Delta\tau}_{\sim T^{-4}} + \alpha_2 \underbrace{\dot{\boldsymbol{\kappa}}^\mathrm{T} \dot{\boldsymbol{\kappa}} \Delta\tau}_{\sim T^{-2}} + \alpha_3 \underbrace{\dot{\boldsymbol{a}}_\mathrm{t}^\mathrm{T} \dot{\boldsymbol{a}}_\mathrm{t} \Delta\tau}_{\sim T^{-6}} + \alpha_4 \underbrace{\boldsymbol{v}^\mathrm{T} \mathrm{sign}(\boldsymbol{v}) T \Delta\tau}_{\sim T^0} + \alpha_5 T^2. \quad (5.8)$$

The dependencies of the respective terms on the parameter $T$ are either parabolic, hyperbolic or vanish, but all in even powers of $T$. Therefore, the profile of $J$ with respect to the $T$-$J$-plane is symmetric with respect to the abscissa and has two minima, see Fig. 5.1. For plausibility reasons a positive $T$ is required, which can be ensured by choosing a positive starting value and carefully controlling the convergence behaviour of the optimisation.

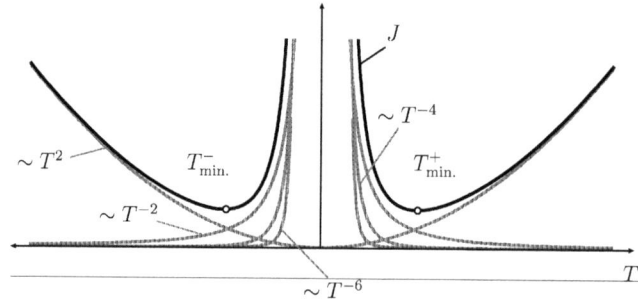

Figure 5.1: Profile of the performance functional with respect to the $T$-$J$-plane.

The gradient of the performance functional $J$ with respect to the optimisation parameters $\boldsymbol{p}$ is given by

$$\frac{\partial J}{\partial \boldsymbol{p}_v} = \underbrace{4\alpha_1 (\boldsymbol{a}_\mathrm{n} \circ \boldsymbol{v} \circ \boldsymbol{\kappa}) \boldsymbol{F}_v^\mathrm{T} \frac{\Delta\tau}{T} + 2\alpha_3 \dot{\boldsymbol{a}}_\mathrm{t} \boldsymbol{F}_v''^\mathrm{T} \frac{\Delta\tau}{T^3} \alpha_4 \mathrm{sign}(\boldsymbol{v}) \boldsymbol{F}_v^\mathrm{T} \Delta\tau,}_{[n_v \times 1]} \quad (5.9)$$

## 5.4. TRAJECTORY GENERATION

$$\frac{\partial J}{\partial \boldsymbol{p}_\kappa} = \underbrace{2\alpha_1(\boldsymbol{a}_\mathrm{n}\circ\boldsymbol{v}\circ\boldsymbol{v})\boldsymbol{F}_\kappa^\mathrm{T}\Delta\tau + 2\alpha_2\dot{\boldsymbol{\kappa}}\boldsymbol{F}_\kappa^{\prime\mathrm{T}}\frac{\Delta\tau}{T}}_{[n_\kappa\times 1]}, \qquad (5.10)$$

and

$$\frac{\partial J}{\partial T} = \underbrace{-4\alpha_1\boldsymbol{a}_\mathrm{n}^\mathrm{T}\boldsymbol{a}_\mathrm{n}\frac{\Delta\tau}{T} - 2\alpha_2\dot{\boldsymbol{\kappa}}^\mathrm{T}\dot{\boldsymbol{\kappa}}\frac{\Delta\tau}{T} - 6\alpha_3\dot{\boldsymbol{a}}_\mathrm{t}^\mathrm{T}\dot{\boldsymbol{a}}_\mathrm{t}\frac{\Delta\tau}{T} + 2\alpha_5 T}_{[1\times 1]}, \qquad (5.11)$$

and combined

$$\boldsymbol{\nabla}\boldsymbol{J}^\mathrm{T} = \frac{\partial J}{\partial \boldsymbol{p}}^\mathrm{T} = \underbrace{\left[\frac{\partial J}{\partial \boldsymbol{p}_v}^\mathrm{T}\; \frac{\partial J}{\partial \boldsymbol{p}_\kappa}^\mathrm{T}\; \frac{\partial J}{\partial T}^\mathrm{T}\right]^\mathrm{T}}_{[1\times n]}. \qquad (5.12)$$

### 5.4.3 Obstacle avoidance

Only those obstacle edges, represented by the coordinates of their delimiting corner points (Sec. 5.5), that are visible from the starting point of the trajectory, are considered. Furthermore, as a consequence of how the waypoints are selected (Sec. 5.6) it can be assumed, that no obstacles intersect the direct connection between starting point $\boldsymbol{0}$ and target $\boldsymbol{X}_1$, i.e. the target is visible from the starting point.
To determine obstacle collision, each visible obstacle edge is classified, see Fig. 5.2,

1. Facing $\overline{\boldsymbol{0X}_1}$: For each node all four corners of the assumed rectangular vehicle are checked for their normal distance from the obstacle edge $\overline{\boldsymbol{H}_1\boldsymbol{H}_2}$. To the current obstacle edge, the minimal distance $\delta_\mathrm{min}$ and the index $i_\mathrm{min}$ of the corresponding node is assigned. A corrected position $\boldsymbol{X}_\mathrm{des}$ for the node $i_\mathrm{min}$ is obtained by moving it perpendicularly towards $\overline{\boldsymbol{0X}_1}$.

2. Facing away from $\overline{\boldsymbol{0X}_1}$: Each connection between two subsequent nodes is checked for intersection with $\overline{\boldsymbol{H}_1\boldsymbol{H}_2}$. If there is an intersection, $\delta_\mathrm{min}$ is determined as the projection of the vector from the node before the intersection to the obstacle point $\boldsymbol{H}_1$ or $\boldsymbol{H}_2$ next to $\overline{\boldsymbol{0X}_1}$ on the tangential direction of the obstacle. $\boldsymbol{X}_\mathrm{des}$ is obtained by moving the node $i_\mathrm{min}$ along the obstacle edge towards $\overline{\boldsymbol{0X}_1}$.

If the smallest $\delta_\mathrm{min}$ encountered is negative, i.e. there is a collision with at least one obstacle, the constraints are activated for the node with index $i_\mathrm{min}$, Sec. 5.4.4.

### 5.4.4 Constraints

A number of equality and inequality constraints can be respected by the proposed algorithm:

1. Initial constraints at starting point $A$ to enable jerk-free transition between two trajectory sections: Velocity and curvature to ensure continuous lateral acceleration, (5.4), and continuous tangential acceleration. Continuous curvature by itself is required so as to prevent sudden re-alignment of the wheels.

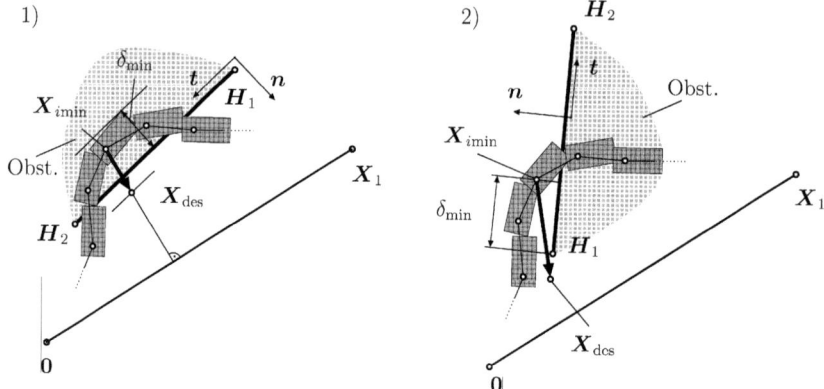

Figure 5.2: Obstacle avoidance.

2. Target ($B$) position, orientation, velocity and acceleration.

3. Inequality constraints: Curvature within bounds given by maximum steering angle. Maximum velocity and maximum acceleration given by the vehicle's dynamic specifications.

4. Position constraints for obstacle avoidance.

The collected constraints are given by

$$\boldsymbol{G}^{\mathrm{T}} = \begin{bmatrix} \left. \begin{array}{c} v_1 - v_A \\ a_{t,1} - a_{t,A} \\ \kappa_1 - \kappa_A \end{array} \right\} \ 1) \\ \left. \begin{array}{c} \varphi_m - \varphi_B \\ x_m - x_B \\ y_m - y_B \\ v_m - v_B \\ a_{t,m} - a_{t,B} \end{array} \right\} \ 2) \\ \left. \begin{array}{c} \boldsymbol{\kappa}_{\max} - \mathrm{sign}(\boldsymbol{\kappa}_{\max})\kappa_{\lim} \\ \boldsymbol{v}_{\max} - \mathrm{sign}(\boldsymbol{v}_{\max})v_{\lim} \\ \boldsymbol{a}_{t,\max} - \mathrm{sign}(\boldsymbol{a}_{t,\max})a_{t,\lim} \end{array} \right\} \ 3) \\ \left. \begin{array}{c} x_{i\min} - x_{\mathrm{des}} \\ y_{i\min} - y_{\mathrm{des}} \end{array} \right\} \ 4) \end{bmatrix} = \boldsymbol{0}, \quad (5.13)$$

where the inequality constraints for the first at most $l$ nodes at which the constraint is exceeded and position constraints for obstacle avoidance are activated when appropriate, i.e. when an inequality or obstacle is violated, respectively. Otherwise they are inactive.
The gradients of initial velocity and acceleration are easily obtained as

$$\frac{\partial G_1}{\partial \boldsymbol{p}} = \frac{\partial v_1}{\partial \boldsymbol{p}} = \begin{bmatrix} \boldsymbol{F}_{v,1} \frac{1}{T} \\ 0 \end{bmatrix} \quad (5.14)$$

## 5.4. TRAJECTORY GENERATION

and

$$\frac{\partial G_2}{\partial \boldsymbol{p}} = \frac{\partial a_{t,1}}{\partial \boldsymbol{p}} = \begin{bmatrix} \boldsymbol{F}'_{v,1}\frac{1}{T^2} \\ \boldsymbol{0} \end{bmatrix}, \tag{5.15}$$

where the derivatives with respect to $T$, exhibiting negative powers of $T$, are deliberately replaced by zero. Otherwise the algorithm would try to fulfil the constraints by increasing $T$ beyond all bounds, when the reference values $(v_A, a_{t,A})$ are equal to zero. This effect, however, collides with the result of the optimisation and is not desirable.

Also the gradient of the initial curvature is readily calculated as

$$\frac{\partial G_3}{\partial \boldsymbol{p}} = \frac{\partial \kappa_1}{\partial \boldsymbol{p}} = \begin{bmatrix} \boldsymbol{0} \\ \boldsymbol{F}_{\kappa,1} \\ 0 \end{bmatrix}. \tag{5.16}$$

The gradient of the terminal orientation reads

$$\frac{\partial G_4}{\partial \boldsymbol{p}} = \frac{\partial \varphi_m}{\partial \boldsymbol{p}} = \begin{bmatrix} \boldsymbol{F}_{v,m-1}\boldsymbol{\kappa}_{m-1}\Delta\tau \\ \boldsymbol{F}_{\kappa,m-1}\boldsymbol{v}_{m-1}T\Delta\tau \\ 0 \end{bmatrix}, \tag{5.17}$$

while the derivation of the gradients of the terminal coordinates is slightly more involved since it contains a double discrete integration,

$$\frac{\partial G_5}{\partial \boldsymbol{p}} = \frac{\partial x_m}{\partial \boldsymbol{p}} = \begin{bmatrix} \boldsymbol{F}_{v,m-1}\cos(\boldsymbol{\varphi}_{m-1})\Delta\tau - \frac{\partial \boldsymbol{\varphi}_{m-1}}{\partial \boldsymbol{p}_v}(\boldsymbol{v}_{m-1}\circ\sin\boldsymbol{\varphi}_{m-1})T\Delta\tau \\ -\frac{\partial \boldsymbol{\varphi}_{m-1}}{\partial \boldsymbol{p}_\kappa}(\boldsymbol{v}_{m-1}\circ\sin\boldsymbol{\varphi}_{m-1})T\Delta\tau \\ 0 \end{bmatrix} \tag{5.18}$$

and

$$\frac{\partial G_6}{\partial \boldsymbol{p}} = \frac{\partial y_m}{\partial \boldsymbol{p}} = \begin{bmatrix} \boldsymbol{F}_{v,m-1}\sin(\boldsymbol{\varphi}_{m-1})\Delta\tau + \frac{\partial \boldsymbol{\varphi}_{m-1}}{\partial \boldsymbol{p}_v}(\boldsymbol{v}_{m-1}\circ\cos\boldsymbol{\varphi}_{m-1})T\Delta\tau \\ \frac{\partial \boldsymbol{\varphi}_{m-1}}{\partial \boldsymbol{p}_\kappa}(\boldsymbol{v}_{m-1}\circ\cos\boldsymbol{\varphi}_{m-1})T\Delta\tau \\ 0 \end{bmatrix}, \tag{5.19}$$

where a vector (in bold) with index $i$ denotes the full vector truncated after element $i$ and a matrix with index $i$ denotes the full matrix with every row vector truncated after element $i$. The derivatives of the orientations are given by the matrices

$$\frac{\partial \boldsymbol{\varphi}_{m-1}}{\partial \boldsymbol{p}_v} = \underbrace{\begin{bmatrix} \boldsymbol{0}, & \kappa_1 \boldsymbol{F}_{v,1}\Delta\tau, & \ldots, & \sum_{k=1}^{m-2} \kappa_k \boldsymbol{F}_{v,k}\Delta\tau \end{bmatrix}}_{[n_v \times m-1]} \tag{5.20}$$

and

$$\frac{\partial \boldsymbol{\varphi}_{m-1}}{\partial \boldsymbol{p}_\kappa} = \underbrace{\begin{bmatrix} \boldsymbol{0}, & v_1 \boldsymbol{F}_{\kappa,1}T\Delta\tau, & \ldots, & \sum_{k=1}^{m-2} v_k \boldsymbol{F}_{\kappa,k}T\Delta\tau \end{bmatrix}}_{[n_\kappa \times m-1]}. \tag{5.21}$$

The terminal velocity and acceleration are treated in the same way as the initial ones,

$$\frac{\partial G_7}{\partial p} = \frac{\partial v_m}{\partial p} = \begin{bmatrix} F_{v,m}\frac{1}{T} \\ 0 \end{bmatrix} \qquad (5.22)$$

and

$$\frac{\partial G_8}{\partial p} = \frac{\partial a_{t,m}}{\partial p} = \begin{bmatrix} F'_{v,m}\frac{1}{T^2} \\ 0 \end{bmatrix}. \qquad (5.23)$$

The inequality constraints' derivatives read

$$\frac{\partial G_9}{\partial p} = \frac{\partial \kappa_{\max}}{\partial p} = \underbrace{\begin{bmatrix} 0 \\ F_{\kappa,\max} \\ 0 \end{bmatrix}}_{[n \times l]}, \qquad (5.24)$$

$$\frac{\partial G_{10}}{\partial p} = \frac{\partial v_{\max}}{\partial p} = \underbrace{\begin{bmatrix} F_{v,\max}\frac{1}{T} \\ 0 \\ -\frac{1}{T}v_{\max} \end{bmatrix}}_{[n \times l]} \qquad (5.25)$$

and

$$\frac{\partial G_{11}}{\partial p} = \frac{\partial a_{t,\max}}{\partial p} = \underbrace{\begin{bmatrix} F'_{v,\max}\frac{1}{T^2} \\ 0 \\ -\frac{1}{T}a_{t,\max} \end{bmatrix}}_{[n \times l]}. \qquad (5.26)$$

In this case the time-derivatives do not need to be replaced by zero, because the reference values ($v_{\text{lim}}$ and $a_{t,\text{lim}}$) are obviously non-zero by definition.

The gradients of the position constraints for obstacle avoidance are calculated analogously to (5.18) and (5.19).

The full gradient matrix of all active constraints is given by

$$\boldsymbol{\nabla G} = \underbrace{\begin{bmatrix} \frac{\partial G_1}{\partial p}, \ldots, \frac{\partial G_o}{\partial p} \end{bmatrix}}_{[n \times o]}, \qquad (5.27)$$

where $o$ denotes the indices of the active constraints.

## 5.4.5 Calculation of the parameter update

In Fig. 5.3 the principle is illustrated in two dimensions: In every step $\nu$ of the iteration the squared Euclidean distance between the negative scaled gradient $-\boldsymbol{\delta}_1 \circ \boldsymbol{\nabla} J$ of the performance functional $J$ and the parameter update $\boldsymbol{\Delta p}$ is minimised while simultaneously approaching the subspace defined by $\boldsymbol{G} = \boldsymbol{0}$.

This procedure could be classified as a constrained gradient method with adaptive step width. The problem is formulated as

## 5.4. TRAJECTORY GENERATION

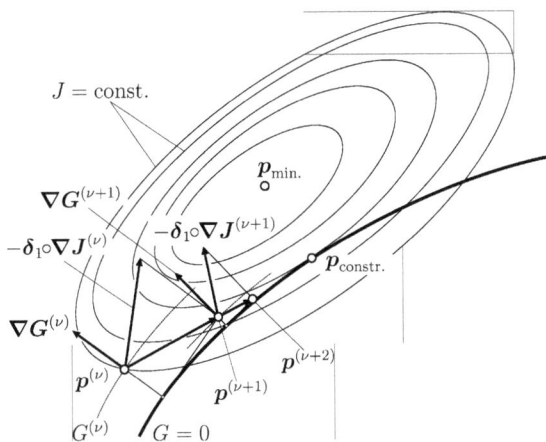

Figure 5.3: Nonlinear constrained optimisation.

$$(\Delta p + \delta_1 \circ \nabla J)^{\mathrm{T}}(\Delta p + \delta_1 \circ \nabla J) \to \min_{\Delta p} \quad (5.28)$$
$$G^{\mathrm{T}} + \nabla G^{\mathrm{T}} \Delta p \stackrel{!}{=} 0,$$

where $\delta_1$ denotes the $n$-vector of positive scaling factors.

To ensure that (5.28) is well-posed, the relative condition number of $\nabla G^{\mathrm{T}} \nabla G$ is checked (naturally, this also implies that the number of constraints must not exceed the number of parameters). If it falls below a threshold $r_{\mathrm{cond,min}}$, a singular value decomposition of $\nabla G$ is calculated according to

$$USV^{\mathrm{T}} = \nabla G^{\mathrm{T}}. \quad (5.29)$$

The columns of $U$ corresponding to the singular values $\sigma$ (ordered by descending magnitude along the main diagonal of the square part of $S$) fulfilling $\sigma^2/\sigma_{\max}^2 < r_{\mathrm{cond,min}}$ are removed, thus obtaining a $o \times \tilde{o}$-matrix $\tilde{U}$. The $o$ constraint equations are then left-multiplied with $\tilde{U}^{\mathrm{T}}$. Thus the number of equations is reduced to $\tilde{o}$ equations which are linearly independent. A linear combination of the constraint equations defined by $\tilde{U}$ results, which means that either a trade-off between contradictory constraints is formulated, or redundant constraints are merged.

Forming a scalar Lagrange-function $\mathcal{L}(\Delta p, \lambda)$

$$\mathcal{L}(\Delta p, \lambda) = (\Delta p + \delta_1 \circ \nabla J)^{\mathrm{T}}(\Delta p + \delta_1 \circ \nabla J) + 2\lambda^{\mathrm{T}}(\tilde{U}^{\mathrm{T}} G^{\mathrm{T}} + \tilde{U}^{\mathrm{T}} \nabla G^{\mathrm{T}} \Delta p) \quad (5.30)$$

and setting its derivatives with respect to $\Delta p$ and $\lambda$ equal to zero

$$\frac{\partial \mathcal{L}}{\partial \Delta p} = 2\Delta p + 2\delta_1 \circ \nabla J + 2\nabla G \tilde{U} \lambda = 0$$
$$\frac{\partial \mathcal{L}}{\partial \lambda} = 2\tilde{U}^T G^T + 2\tilde{U}^T \nabla G^T \Delta p = 0 \tag{5.31}$$

yields the full-rank system of equations of dimension $n + \tilde{o}$

$$\begin{bmatrix} I & \nabla G \tilde{U} \\ \tilde{U}^T \nabla G^T & 0 \end{bmatrix} \begin{bmatrix} \Delta p \\ \lambda \end{bmatrix} = \begin{bmatrix} -\delta_1 \circ \nabla J \\ -\tilde{U}^T G^T \end{bmatrix}, \tag{5.32}$$

which is solved for $\Delta p$ and $\lambda$ during every step of the iteration.

The scaling factors $\delta_1$ are determined by a trade-off between potential reduction of the performance value and Euclidean norm of the parameter update. The approach is similar to Levenberg-Marquardt, but aims at adapting the step width to enhance convergence behaviour. The principle is outlined in the following: A local linearisation of a performance functional is given by

$$J^{(\nu+1)} \doteq J^{(\nu)} + \nabla J^T \Delta p = J^{(\nu)} - \nabla J^T \nabla J \delta_1, \tag{5.33}$$

and the aforementioned trade-off, employing a weighting factor $\alpha$ is formulated by

$$J^{(\nu+1)\,2} + \alpha \Delta p^T \Delta p = J^{(\nu+1)\,2} + \alpha \nabla J^T \nabla J \delta_1^2 \to \min_{\delta_1}. \tag{5.34}$$

Inserting (5.33) into (5.34), differentiating with respect to $\delta_1$, setting the derivative equal to zero and solving for $\delta_1$ leads to

$$\delta_1 = \frac{J^{(\nu)}}{\nabla J^T \nabla J + \alpha}. \tag{5.35}$$

The thereby calculated variation of the step width is superposed on different basic step widths for the different groups of parameters to form the $n$-vector $\boldsymbol{\delta}_1$

$$\boldsymbol{\delta}_1 = \frac{J}{\|\nabla J\|_2^2 + \alpha} \begin{bmatrix} \delta_{1,v} \mathbf{1} \\ \delta_{1,\kappa} \mathbf{1} \\ \delta_{1,T} \end{bmatrix}. \tag{5.36}$$

Finally, the parameters are updated using a relaxation factor $\delta_2$ for convergence control

$$\begin{bmatrix} p_v \\ p_\kappa \\ T \end{bmatrix}^{(\nu+1)} = \begin{bmatrix} p_v \\ p_\kappa \\ T \end{bmatrix}^{(\nu)} + \delta_2 \Delta p, \tag{5.37}$$

and the descriptive quantities (5.3) through (5.6) are re-calculated.

After a specified number of iterations, or when a combined criterion consisting of the Euclidean of the parameter update and the mean absolute constraint residuum fulfils

## 5.4. TRAJECTORY GENERATION

$$\|\mathbf{\Delta p}\|_2 + \frac{1}{o}\sum_{k=1}^{o}|G_k| < c, \qquad (5.38)$$

the algorithm terminates.

### 5.4.6 Tuning parameters

The relevant parameters used to tune the algorithm are

- Number of nodes $m$: Large enough to satisfy accuracy requirements, but as small as possible to save calculation time.

- Number of basis functions $n_v + n_\kappa$: Large enough to ensure sufficient flexibility in difficult environments, but as small as possible to save calculation time.

- Maximum number of active inequality constraints $l$: Large enough to enforce accurate fulfilment of the inequalities, as small as possible with regard to calculation time, but in any case small enough to ensure that the total number of constraints does not exceed the number of parameters.

- Regularisation parameter $\alpha$, step widths $\delta_{1,v}$, $\delta_{1,\kappa}$, $\delta_{1,T}$ and $\delta_2$: Small enough to avoid oscillatory behaviour, but as large as possible to achieve fast convergence.

- Number of iterations performed or termination criterion $c$: Determine trade-off between accuracy and calculation time.

- Initial values for the parameters: Their suitable choice can significantly influence the calculation time.

- Weighting factors $\alpha_1$ through $\alpha_5$: Determine the shape of the optimised trajectory.

### 5.4.7 Post-processing for trajectory tracking control

To be fed to a trajectory tracking controller working at a constant sampling time $T_s$, the resulting trajectory is resampled. Therefore, the original total time $T$ is slightly modified to obtain an integer number of sampling intervals,

$$\mathrm{d}\tilde{\tau} = T_s/\tilde{T} = 1/\mathrm{round}(T/T_s). \qquad (5.39)$$

With the definitions $\underline{x}_i := x(\mathrm{floor}(\frac{\Delta\tilde{\tau}}{\Delta\tau}(i-1))+1)$ and $\overline{x}_i := x(\mathrm{ceil}(\frac{\Delta\tilde{\tau}}{\Delta\tau}(i-1))+1)$ a linear interpolation of the cartesian trajectory coordinates and the orientation is formulated, exemplarily written for the $x$-coordinate

$$\tilde{x}_i = \underline{x}_i + \left(\overline{x}_i - \underline{x}_i\right)\left(\frac{\Delta\tilde{\tau}}{\Delta\tau}(i-1) - \mathrm{floor}\left(\frac{\Delta\tilde{\tau}}{\Delta\tau}(i-1)\right)\right). \qquad (5.40)$$

## 5.5 Map-building

As stated in the introduction, it is assumed that obstacle edges are represented by straight lines as a result of some suitable preprocessing of laser or ultrasonic range data, stereovision measurement, etc. At each sampling instant currently visible obstacle edges $L_{vis}$ (given by $H_1$ and $H_2$ in order of ascending angle, i.e. from right to left) are merged with previously available information $L$, thus gradually building a 2D-map of the environment.

```
FOR all L_vis (index i)
    flag = 0
    FOR all L (index j)
        IF flag == 0: L_vis,i overlaps L_j for the first time
            IF collinear
                IF H_{2,j} ∈ H_1 H_{2i}
                    overwrite H_{2,j} with H_{2,i}
                    flag = 1
                    index = j
                IF H_{1,j} ∈ H_1 H_{2i}
                    overwrite H_{1,j} with H_{1,i}
                    flag = 1
                    index = j
                IF H_1 H_{2i} ∈ H_1 H_{2j}
                    don't overwrite
                    flag = 1
                    index = j
        ELSE: L_vis,i has previously been merged
            IF collinear
                IF H_{2,index} ∈ H_1 H_{2j}
                    overwrite H_{2,index} with H_{2,j}
                    mark j for deletion
                IF H_{1,index} ∈ H_1 H_{2j}
                    overwrite H_{1,index} with H_{1,j}
                    mark j for deletion
                IF H_1 H_{2j} ∈ H_1 H_{2index}
                    don't overwrite
                    mark j for deletion
    delete obsolete L's if there are any
    IF flag == 0: collinear but no overlap or entirely new
        add L_vis,i to L
```

To check for collinearity and overlap, the points $H_{1,j}$ and $H_{2,j}$ under consideration are expressed in local coordinates $\alpha$ and $\beta$,

$$H_{1,i} = H_{1,2,j} + \alpha t + \beta n \Rightarrow \begin{bmatrix} \alpha \\ \beta \end{bmatrix} = \begin{bmatrix} t & n \end{bmatrix}^{-1} (H_{1,i} - H_{1,2,j}), \tag{5.41}$$

where $t$ and $n$ denote the (not normalised) tangential and normal vectors of obstacle edge $i$.

## 5.6. WAYPOINT GENERATION

If $\beta = 0$ for both $\boldsymbol{H}_{1,j}$ and $\boldsymbol{H}_{2,j}$, the two edges are collinear; if e.g. for $\boldsymbol{H}_{1,j}$ $\alpha \in [0;1]$, $\boldsymbol{H}_{1,j} \in \overline{\boldsymbol{H}_1\boldsymbol{H}_{2i}}$ holds, etc.

## 5.6 Waypoint generation

As long as no direct line of sight to the target exists, a waypoint is sought based on the concept of exterior vertices. From the current position, any visible vertex that is not on both sides connected by visible edges is classified either left or right exterior. The exterior vertices are derived from the currently visible obstacle edges, whereas otherwise the algorithm makes use of the entire environment map. These vertices are considered as the originator for potential waypoints, where a waypoint is found by relocating the vertex by a distance of $d_{\min}$ perpendicularly away from the connection between current position and vertex, see Fig. 5.4.

```
IF target visible
    waypoint = target
ELSE
    candidates = all currently visible exterior vertices
    FOR all candidates (index i)
        IF candidate(i) has been used twice before
            delete candidate(i)
    flag = 0
    WHILE flag<3
        FOR all candidates
            calculate corresponding waypoint, Fig. 5.4
            IF convex or flag == 2
                IF flag == 0
                    check if connection between
                    waypoint and target intersects
                    any edges from front to back
                check normal distance of waypoint
                to all edges
                check Euclidean distance of way-
                point to all vertices
                check visibility of waypoint
                IF all checks o.k.
                    add waypoint to list
                    calculate criterion (5.42)
        IF list empty
            increment flag
        ELSE
            flag = 3: loop terminated
    sort list by criterion
    waypoint = first point in list
    store waypoint to check for multiple occurence
```

The current position is connected with the target via a waypoint, i.e. with two legs. At first, the two legs are required to be convex with respect to the corresponding exterior vertex, i.e. the

vertex is located in the interior of the triangle composed of the two legs and the direct connection from the current position to the target, as shown in Fig. 5.4. If the second leg intersects any known obstacle edge from front to back (exterior to interior), this vertex is discarded. Intersections from back to front are not considered, because in this case it must be assumed that there is a previously unencountered obstacle edge involved. Furthermore, the distance of the candidate waypoint to all edges and its visibility from the current position is checked.
If no vertex fulfils the given conditions, the condition regarding the intersection is dropped. If still no solution can be found, the convexity condition is dropped as well.
If there are multiple solutions, they are ordered according to a criterion containing the deviation angle from the current orientation, the deviation from the straight line to the target and the total length of both legs, see Fig. 5.4,

$$\lambda_1(l_1 + l_2) + \lambda_2|\Delta\varphi| + \lambda_3|\Delta\theta| \to \min, \qquad (5.42)$$

which is tuned by the weighting factors $\lambda_1$ through $\lambda_3$.
Any vertex used as originator for a waypoint is stored, such that it can be excluded from future candidates, if it was already used twice. This measure serves to avoid infinite oscillation between two apparently feasible waypoints, but enables to round a corner where the same vertex is used twice consecutively.

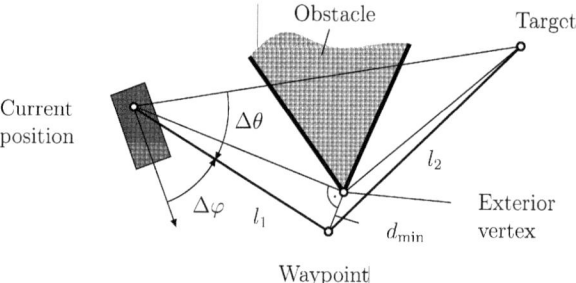

Figure 5.4: Calculation of a waypoint from an exterior vertex.

The desired orientation at the waypoint is normally chosen to be equal to the angle of the direct connection between current position and waypoint, only when a corner is rounded, i.e. the same vertex is chosen twice consecutively, the orientation constraint is dropped. The terminal velocity and acceleration are chosen to be zero to enable jerk-free braking. This choice has proven to be most feasible considering that at the time of waypoint generation no reliable knowledge about the direction of motion beyond the waypoint is available.

# Chapter 6

# Results

To demonstrate the performance of the proposed navigation and control algorithm, two different trajectories to be tracked have been generated:

- A corner with linear acceleration and deceleration and a cosine-shaped curvature profile, thus producing large side-accelerations.
- A square with sinusoidal velocity and yaw rate profiles, where the robot turns on the spot at the vertices, demanding high translative accelerations and decelerations.

In Fig. 6.1 for both trajectories the reference and actual sequence of configurations are depicted.

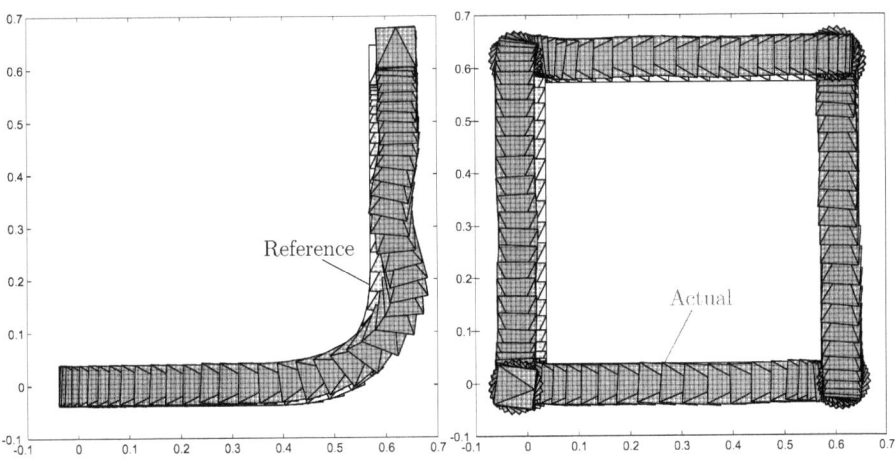

Figure 6.1: Reference and actual trajectories (obtained by video measurement) for corner (left) and square (right).

## 6.1 Trajectory tracking

The performance of the predictive tracking control algorithm is demonstrated by comparing the modified reference trajectory $^C\boldsymbol{x}_{\text{ref}}$ to the measured posture.

In Fig. 6.2 the effect of side-slip in the tracking control algorithm can be observed: the estimated side-slip angle $\alpha$ is subtracted from the reference path angle, thus improving the compensation of the lateral position error $p_\text{n}$. A transient deviation, however, cannot entirely be eliminated.

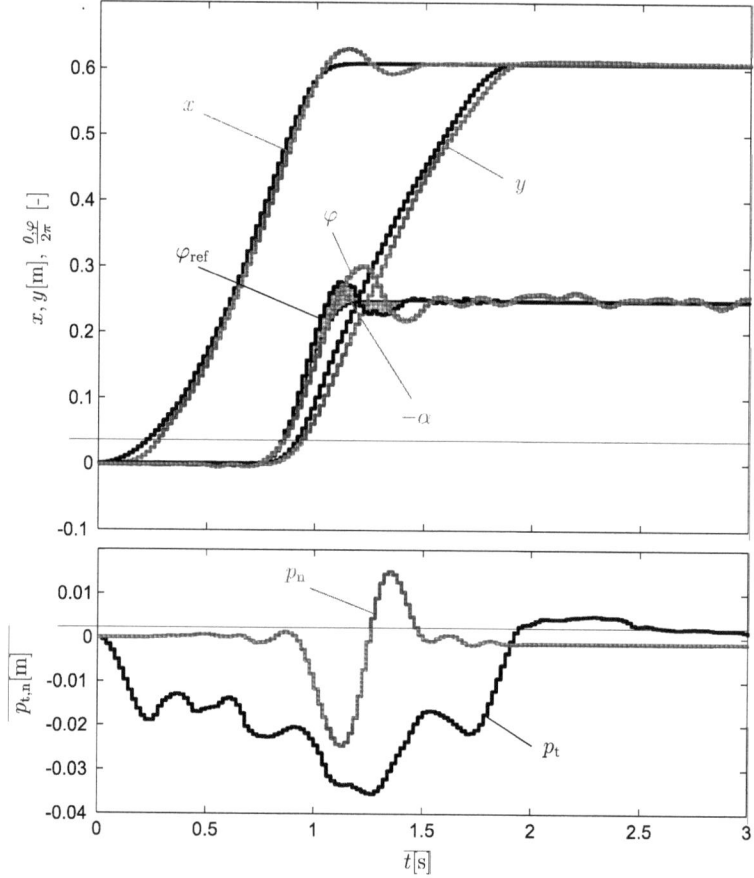

Figure 6.2: Corner-shaped trajectory: Reference posture after slip control $^C\boldsymbol{x}_{\text{ref}}$ (black) vs. actual posture (grey) (top) and tangential and lateral control position error (bottom).

In Fig. 6.3 (top) the compensation of lateral position errors is pointed out by the two zoom-

## 6.2. POINT STABILISATION

windows. This effect can also be observed in Fig. 6.3 (bottom), where a lateral position error after turning on the spot is compensated while moving along the sides of the square (a). The tangential position error at the end of the trajectory also converges towards zero (b).

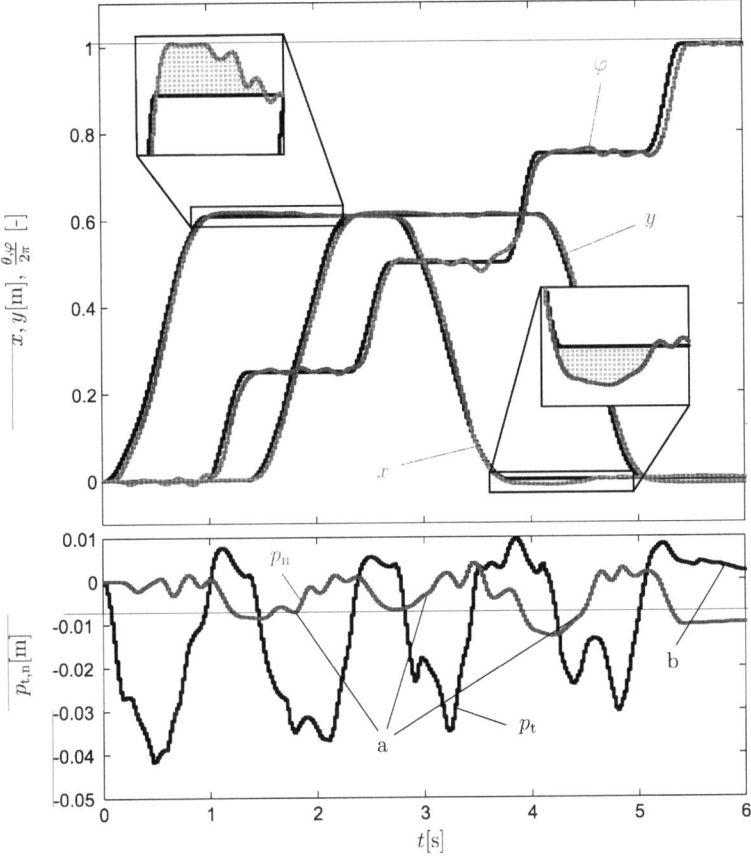

Figure 6.3: Square-shaped trajectory: Reference posture after slip control $^{C}\boldsymbol{x}_{\text{ref}}$ (black) vs. actual posture (grey) (top) and tangential and lateral control position error (bottom).

## 6.2 Point stabilisation

Compensation of a lateral position error at standstill is demonstrated by prescribing a stationary point in lateral direction of the robot as a reference trajectory. This is an extreme situation

designed for testing purposes. In practice, trajectory tracking will not usually leave the robot with a residual lateral posture error of such magnitude.
In Fig. 6.4 the measured postures are compared to the reference posture to demonstrate the action of the control algorithm.

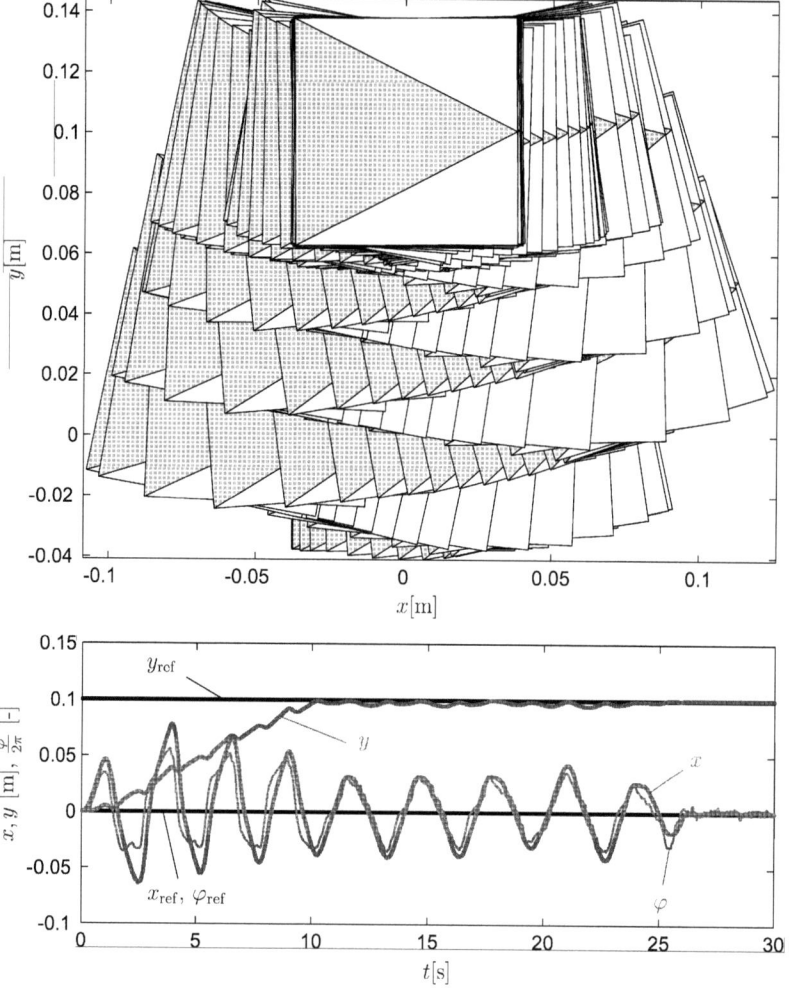

Figure 6.4: Sequence of configurations (top) and reference and measured posture (bottom) while compensating a lateral position error at standstill.

## 6.3 Navigation

The navigation algorithm was evaluated using the video measurement system described in Sec. 2.4.1. Its data is processed offline (i.e. after the experiment) and compared with logged navigation data obtained via the RF-module from the autonomous robot, see Appendix A.

In Figs. 6.5 and 6.6 the effect of the slip control algorithm is obvious during tracking of the corner. Due to high side-acceleration, a side-slip angle is built up, which is correctly estimated. The slip control algorithm reduces the step size, resulting in a delay of around five samples. Subsequently, when circumstances are safe with regard to slip, the step size is increased to catch up with the original trajectory.

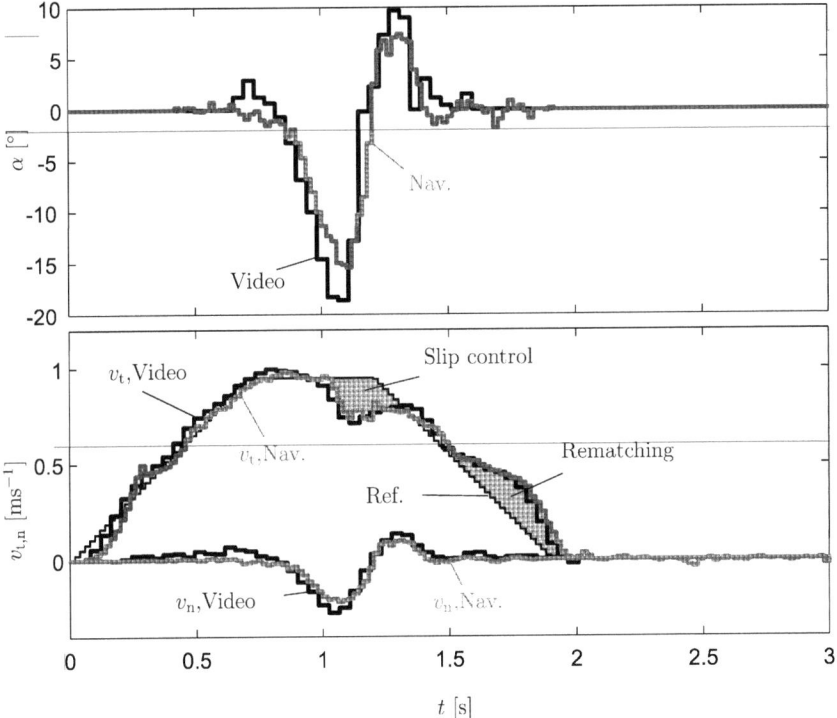

Figure 6.5: Side slip angle estimate (top), tangential and lateral velocity estimate (bottom) during tracking of the corner. Ref. indicates the velocities associated with the reference trajectory.

In Fig. 6.7 the action of the slip control algorithm for tangential wheel slip is demonstrated for tracking of the square. Whenever tangential slip is detected (indicated by the black circles),

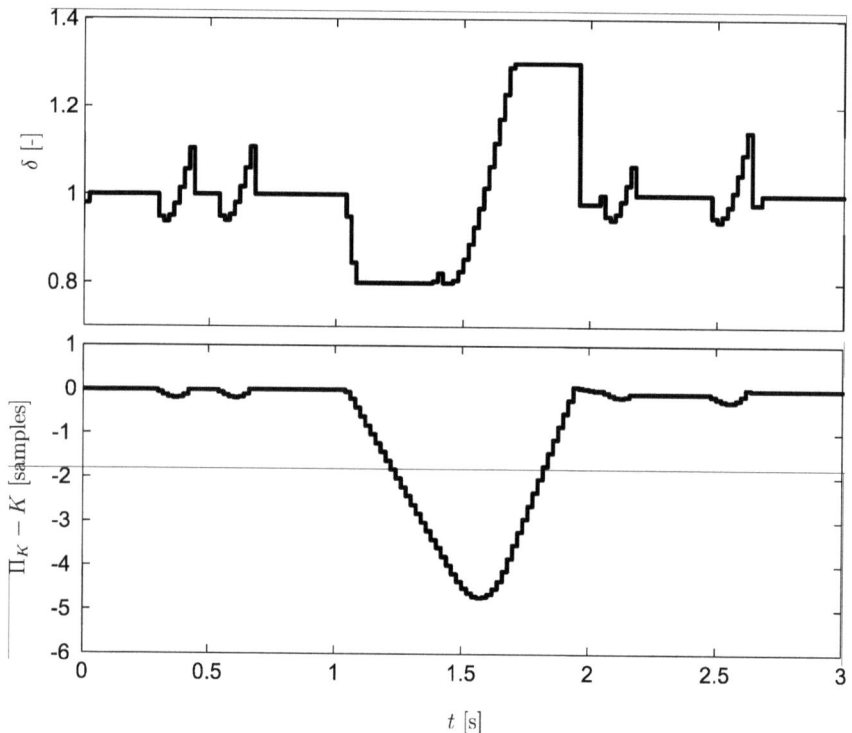

Figure 6.6: Step size (top) and delay (bottom) during tracking of the corner.

the step size is temporarily reduced. In these situations, also a temporary divergence of the velocity estimate can be observed (indicated by the light grey areas), which is a consequence of the reaction time of switching between odometric and inertial estimation.

## 6.3. NAVIGATION

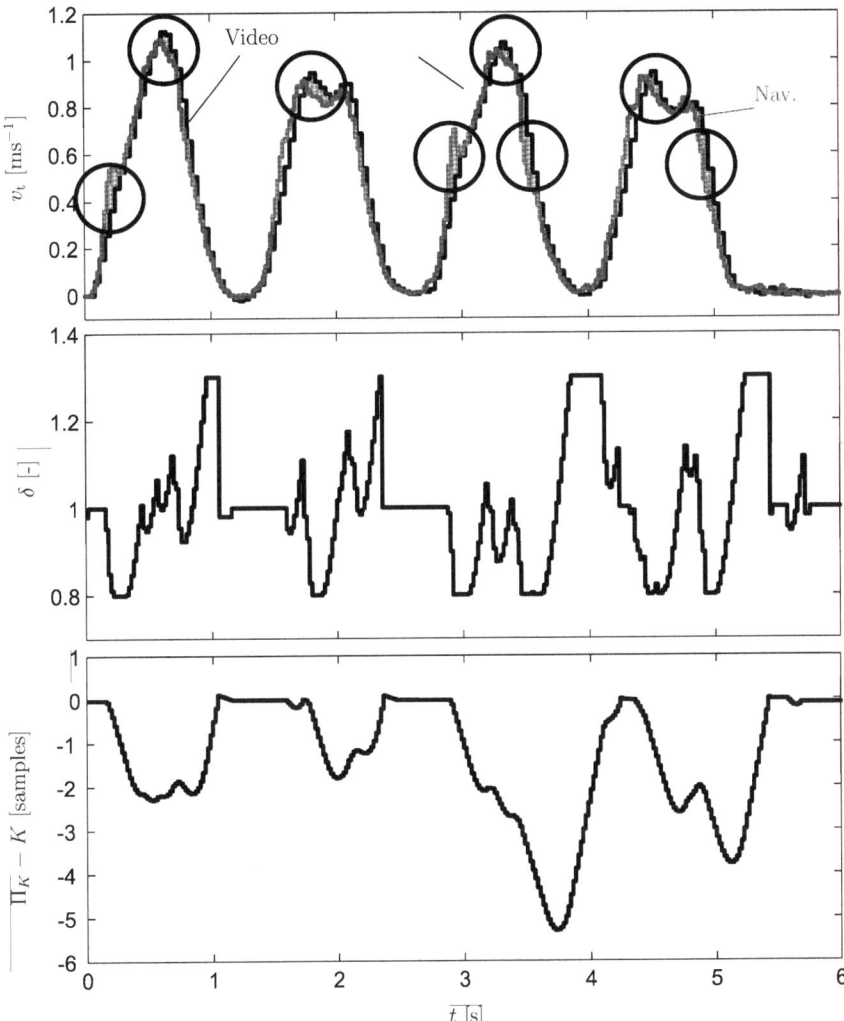

Figure 6.7: Tangential velocity estimation (top), step size (middle) and delay (bottom) during tracking of a square.

In Figs. 6.8 and 6.9 the absolute navigation position error (Euclidean distance between reference position and actual position) is depicted for both trajectories. It can be observed that the navigation errors increase over time, which is inevitable in proprioceptive navigation, as stated in Sec. 4.1. The final navigation error relating to the geometric length of the trajectory is 2.5% for the corner and 1% for the square.

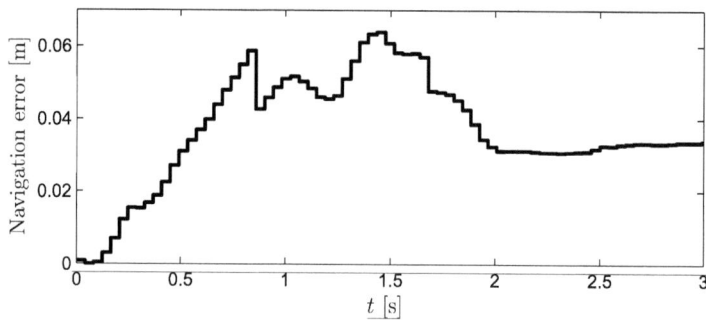

Figure 6.8: Navigation error during tracking of the corner-shaped trajectory.

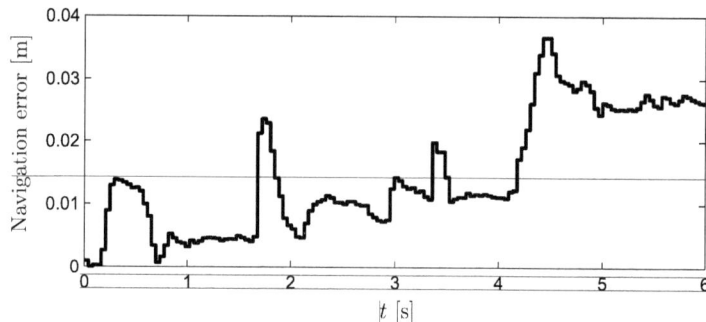

Figure 6.9: Navigation error during tracking of the square-shaped trajectory.

## 6.4 Motion planning

To demonstrate the performance of the proposed motion planning algorithm, enclosed environments cluttered with randomly distributed and shaped triangles have been generated.
In Fig. 6.10 an overview of map-building and waypoint generation is given. For this environment, five intermediate waypoints are required, resulting in six trajectory sections.

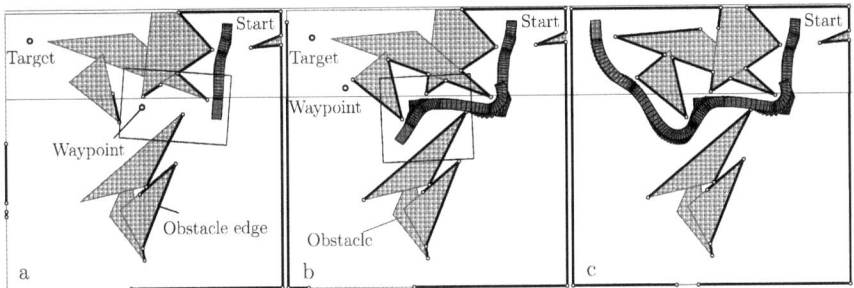

Figure 6.10: Mapbuilding and waypoint selection (in global coordinates): The solid black lines indicate the map of the environment at each stage (a-c).

In Fig. 6.11 three planning-stages during optimisation of the trajectory are depicted, showing how the trajectory is gradually altered to stay clear of an obstacle.

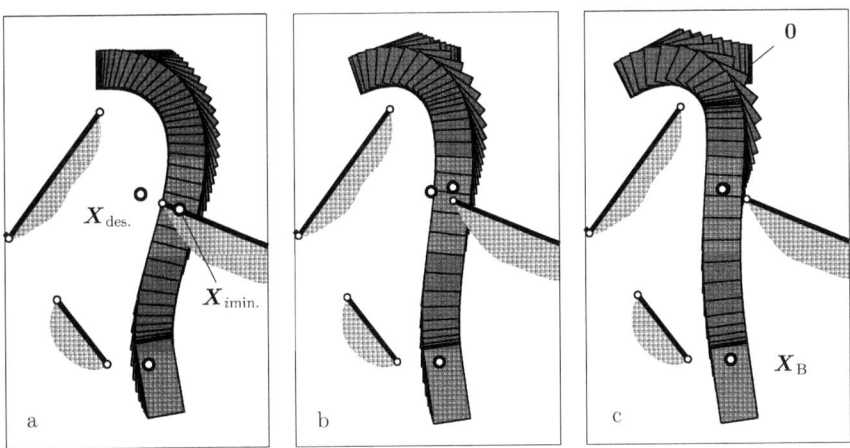

Figure 6.11: Obstacle avoidance for the trajectory in the region indicated by the rectangle in Fig. 6.10(a): From a-c the trajectory is altered to stay clear of the obstacle, while improving the fulfilment of the terminal position constraint.

In Fig. 6.12 various quantities over time are depicted for the trajectory section Fig. 6.11(c), demonstrating how initial, terminal and inequality constraints are fulfilled.

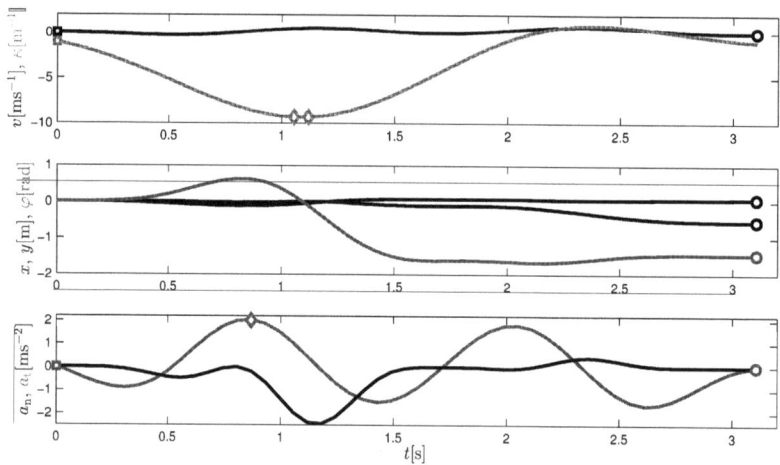

Figure 6.12: Descriptive quantities, initial constraints indicated by a square, terminal constraints by a circle and inequality constraints by a diamond for the trajectory depicted in Fig. 6.11(c).

Finally, in Fig. 6.13 two versions of the section of the trajectory as indicated in Fig. 6.10(b) are plotted to show how different weighting factors produce differently shaped trajectories.

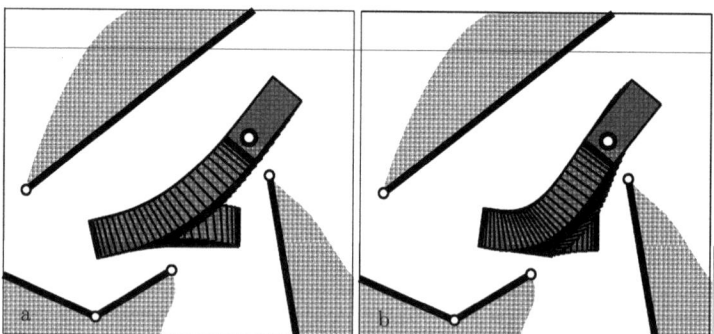

Figure 6.13: Effect of different weighting for the area indicated by the rectangle in Fig. 6.10(b): In (a) the weighting factor for the absolute path length is smaller than in (b), resulting in a smoother but longer path at roughly the same overall time.

# Chapter 7
# Conclusion and outlook

Novel concepts for each of the three sub-tasks motion planning, tracking and navigation in autonomous mobile robot motion control have been presented. The performance of the tracking control algorithm and the navigation algorithm has been practically demonstrated by real-life experiments. The required calculation time for both tracking control and navigation during every sampling interval is well within the bound of 20ms, even though the presented algorithms' complexity exceeds most existing approaches.

As for the motion planning algorithms, for practical evaluation a sensor system for environment perception would be necessary, as well as a more sophisticated map-building algorithm, accounting for localisation and perception uncertainties. Moreover, environment perception would enable simultaneous localisation and map-building (SLAM), thereby introducing an exteroceptive component into navigation and enabling reliable long-term navigation.

The calculation time of the motion planning algorithm amounts to roughly 2s per trajectory section on the desktop PC. Experience from comparing the calculation time of the navigation and control algorithms in simulation with the calculation time in practical implementation on the DSP suggests that the execution time on an embedded system would be of the same order of magnitude. For such a fast system like Tinyphoon this might be too slow, for most practical applications, however, it would be well within the margin of feasibility.

Aside from the missing possibility for environment perception, it is believed that the presented algorithms constitute a solid basis for a practically applicable autonomous mobile robot system.

# Appendix A

# Technical specifications

## A.1 Hardware

### A.1.1 Chassis, transmissions and wheels

The chassis, transmissions, wheels, circuit board and RF-module for PC-communication are designed by Bluetechnix (http://www.tinyphoon.com, http://www.bluetechnix.com) and have been presented by Novak and Mahlknecht in [72]. The relevant parameters are listed in Table A.1.

Table A.1: Mechanical quantities of chassis, wheels and transmissions

| Quantity | Symbol | Value, unit |
|---|---|---|
| Width | $a$ | 0.075m |
| Length | $a$ | 0.075m |
| Wheel radius | $r$ | 0.023m |
| Height of centroid | $s$ | 0.0275m |
| Approx. wheelbase | $b$ | 0.068m |
| Total mass | $m$ | 0.39kg |
| Wheel mass | $m_w$ | 0.02kg |
| Gear ratio | $n$ | $100/12 = 8.33$ |

### A.1.2 DC-motors and incremental encoders

The DC-motors of type 2224SR are manufactured by Faulhaber [34]. The relevant parameters are listed in Table A.2. The electrical quantities do not match the specifications given in the manufacturer's data sheet very well. 2-channel incremental encoders are directly mounted on the DC-motor's shaft, enabling signed velocity measurement.

Due to the finite number of steps per revolution, quantisation noise is introduced into the velocity measurement. The quantisation step for wheel angular velocity measurement in $\text{rads}^{-1}$ is given by

Table A.2: Mechanical and electrical quantities of the DC-motors

| Quantity | Symbol | Value,unit |
|---|---|---|
| Moment of inertia | $I_m$ | $2.7 \cdot 10^{-7}$ kgm$^2$ |
| Resistance | $R$ | $5.8\,\Omega$ |
| Inductance | $L$ | $3.2 \cdot 10^{-5}\,\Omega$s |
| Motor constant | $K$ | $6.42 \cdot 10^{-3}$ Vs or NmA$^{-1}$ |
| Encoder resolution | $N_{\text{Enc}}$ | $4 \times 512$ steps per revolution |
| Quantisation noise variance | $\sigma_{\text{enc}}^2$ | $1.49 \cdot 10^{-6}$ (rads$^{-1}$)$^2$ |

$$q = \frac{2\pi}{N_{\text{Enc}} n T_N}. \tag{A.1}$$

Due to quantisation, any signal value in the interval $\pm \frac{q}{2}$ around a quantisation step is represented by the value of the quantisation step. Therefore, a rectangular probability density function is obtained. The variance of a quantised signal value is obtained by computing the second moment of the probability density function,

$$\sigma_{\text{enc}}^2 = \frac{1}{q} \int_{-q/2}^{q/2} \xi^2 \, \mathrm{d}x = \frac{q^2}{12}. \tag{A.2}$$

### A.1.3 Amplifier

The dual-full-bridge driver L298N is made by STMicroelectronics [88]. The decisive quantities as given in Table A.3 are obtained by measurement.

Table A.3: Electrical quantities of the dual-full-bridge driver

| Quantity | Symbol | Value,unit |
|---|---|---|
| Voltage drop of a transistor | $u_{\text{tr}}$ | 0.6V |
| ~~Voltage drop of a diode~~ | ~~$u_d$~~ | ~~0.35V~~ |
| Switch-on time of a transistor at low | $T_1$ | $1.8 \cdot 10^{-6}$s |
| Switch-off time of a transistor at high | $T_2$ | $1.0 \cdot 10^{-6}$s |
| Switch-off time of a transistor at low | $T_3$ | $0.35 \cdot 10^{-6}$s |

### A.1.4 Acceleration sensors

The robot's acceleration sensors ADXL202 are manufactured by Analog Devices [3]. They are single-chip digital-output piezo-based sensors with two perpendicular channels each. Their output is PWM-encoded, i.e. the PWM-ratio is proportional to the measured accelerations, where zero acceleration corresponds to a PWM-ratio of 0.5. The sensor's characteristics are listed in Table A.4.

## A.1. HARDWARE

Table A.4: Sensor characteristics of the acceleration sensors

| Quantity | Symbol | Value,unit |
|---|---|---|
| Measurement range |  | $\pm 2g = \pm 19.62 \text{ms}^{-2}$ |
| Theoretical PWM period | $T_{acc}$ | 2ms |
| Measured PWM period |  | 1.88ms |
| Theoretical sensitivity | $k_{acc}$ | $12.5\%/g$ |
| Analog bandwidth | $f_{acc}$ | 50Hz |
| Calculated noise variance | $\sigma^2_{acc}$ | $3.08 \cdot 10^{-4} (\text{ms}^{-2})^2$ |
| Measured noise variance | $\sigma^2_{acc}$ | $3.10 \cdot 10^{-4} (\text{ms}^{-2})^2$ |

The sensor's noise variance in $(\text{ms}^{-2})^2$ is calculated according to the data sheet [3] by

$$\sigma^2_{acc} = \left(2 \cdot 10^{-4} \cdot 9.81 \sqrt{1.6 f_{acc}}\right)^2. \tag{A.3}$$

The manufacturer's specifications are validated by evaluating the approximate noise variance of a recorded sensor signal in steady state.

### A.1.5 Gyro sensor

The robot's angular rate sensor ADXRS300 by Analog Devices [4] is a single-chip analog-output resonator gyro. The sensor characteristics are given in Table A.5.

Table A.5: Sensor characteristics of the yaw-rate sensor

| Quantity | Symbol | Value,unit |
|---|---|---|
| Theoretical measurement range | $\pm \omega_{max}$ | $\pm 300°\text{s}^{-1} = \pm 5.23 \text{rads}^{-1}$ |
| True measurement range |  | $\pm 493°\text{s}^{-1} = \pm 8.6 \text{rads}^{-1}$ |
| Analog bandwidth | $f_{yaw}$ | 40Hz |
| Calculated noise variance | $\sigma^2_{yaw}$ | $1.22 \cdot 10^{-4} (\text{rads}^{-1})^2$ |
| Measured noise variance | $\sigma^2_{yaw}$ | $1.48 \cdot 10^{-4} (\text{rads}^{-1})^2$ |

The sensor's noise variance in $(\text{rads}^{-1})^2$ is calculated according to the data sheet [4] by

$$\sigma^2_{yaw} = \left(0.1\pi/180 \sqrt{f_{yaw}}\right)^2. \tag{A.4}$$

### A.1.6 Microcontroller ($\mu$C)

The microcontroller is the widely used Infineon XC167CI [49]. Its capture compare unit (CAP-COM) is used to decode the PWM-encoded acceleration signal and to count the encoder increments while discerning rising and falling edges. The analog-digital converter (ADC) quantises the analog signal from the gyro sensor. The PWM-unit is used to output the PWM-signals to drive the DC-motors. The relevant characteristics are listed in Table A.6.

Table A.6: Characteristics of the $\mu$C

| Quantity | Symbol | Value,unit |
|---|---|---|
| Processor clock rate | | 40MHz |
| CAPCOM counter clock time | $T_{CC}$ | $2 \cdot 10^{-7}$s |
| Theoretical acceleration sensor quantisation step | | 0.0078ms$^{-2}$ |
| Measured acceleration sensor quantisation step | | 0.0098ms$^{-2}$ |
| ADC resolution | $N_{ADC}$ | 10bit = 1024 steps |
| Theoretical gyro sensor quantisation step | | 0.017rads$^{-1}$ |
| ~~Measured gyro sensor quantisation step~~ | | ~~0.017rads$^{-1}$~~ |
| Sampling time for encoder increment counting | $T_N$ | $2 \cdot 10^{-3}$s |
| Quantisation step of the PWM-output signal | | 0.01 |
| Period of the PWM-output signal | $T_{PWM}$ | $5 \cdot 10^{-5}$s |

The theoretical and measured acceleration sensor quantisation steps do not match very well, this is due to differing sensitivity and to the inaccuracy of the acceleration sensor's output PWM-period.
The acceleration quantisation step is calculated by

$$q_{acc} = \frac{100\%}{k_{acc}} \frac{9.81}{T_{acc}/T_{CC}}. \qquad (A.5)$$

The gyro quantisation step is given by

$$q_{enc} = \frac{2\omega_{max}}{N_{ADC}}, \qquad (A.6)$$

if the entire input range of the ADC is exploited.
The original acceleration noise variances are only marginally increased by quantisation and roundoff errors. The difference between the gyro sensor's measured and calculated noise variances is due to the relatively large quantisation step.
The effect of quantisation on the variance of a noisy signal can be verified by a simple simulation.

## A.1.7 Digital signal processor (DSP)

The robot is equipped with an Analog Devices Blackfin BF533 digital signal processor, [2]. The sampling rate of the algorithm executed on the DSP is slaved to the sampling rate of the microcontroller. As soon as a full batch of sensor data has been collected by the microcontroller, the DSP algorithm is triggered. The communication between the microcontroller and the DSP is performed using an SPI-protocol. The implementation of the algorithm is written in C language, using the license-free linear algebra package CLAPACK [6].
The characteristics of the DSP are listed in Table A.7.

## A.1.8 Peripheral devices

All simulation and development work is performed on a 3GHz, 1GB RAM desktop PC, running WindowsXP Professional.

Table A.7: Characteristics of the DSP

| Quantity | Symbol | Value,unit |
|---|---|---|
| Processor clock rate | | 600MHz |
| Sampling rate for the control algorithm | $T_C$ | $2 \cdot 10^{-2}$s |
| SPI data rate | | 1Mbit/s |

The CCD-camera of type A301FC used for video measurement and validation is manufactured by Basler, [11]. It is connected via firewire cable to the desktop PC and is well capable of recording 25 frames per second in colour at a resolution of 640×480 pixels.

Wireless communication with the robot for data logging purposes is established via the robot's built in RF-module and a custom-built USB-RF-device, made by Bluetechnix (http://www.tinyphoon.com, http://www.bluetechnix.com).

Programming and online-debugging of the DSP is performed via the Analog Devices JTAG processor interface, [2].

## A.2 Software

Simulation and development work is primarily done using MATLAB 7.0, [90]. The implementation of the developed algorithms on the DSP is performed using Analog Devices' development environment VisualDSP++ 3.5, [5]. For video capture, the Basler BCAM 1394 V1.8 driver package [12] is used and for the decomposition of the captured video into individual frames the license-free software VirtualDub [96] is applied. Finally, to log sensor data from the robot's microcontroller via RF-module, a custom-built program called Robovis, developed by Bluetechnix (http://www.tinyphoon.com, http://www.bluetechnix.com) is used.

# Appendix B

# The Extended Kalman Filter

## B.1 Analytical derivation

A nonlinear state space system is given by

$$\dot{x} = f(x, u, w), \tag{B.1}$$

where $x \in \mathbb{R}^n$ denotes the state vector, $u \in \mathbb{R}^p$ denotes the input vector, $w \in \mathbb{R}^n$ denotes the system noise and $f(\cdot)$ is a nonlinear map $\mathbb{R}^{2n+p} \to \mathbb{R}^n$. The system noise $w$ is white with covariance matrix $Q$.

The system's discretisation according to the simple forward Euler scheme is given by

$$x_{k+1} = x_k + T f(x_k, u_k, w_k), \tag{B.2}$$

where $k$ denotes the integer sampling instant and $T$ denotes the sampling time. The measurement equation is given by

$$y = g(x, v), \tag{B.3}$$

where $y \in \mathbb{R}^m$ is the measurement vector, $v \in \mathbb{R}^m$ the white measurement noise with covariance matrix $R$ and $g(\cdot)$ a possibly nonlinear map $\mathbb{R}^n \to \mathbb{R}^m$. Its discretisation is trivially obtained by

$$y_k = g(x_k, v_k). \tag{B.4}$$

The two-stage Extended Kalman Filter estimates the states of the system in two steps. First, the states are predicted using the discrete nonlinear representation (B.2). Second, the predicted states are updated using the difference between the predicted measurements and the actual measurements under minimisation of the covariance of the updated states.

The states at instant $k+1$, based on data of instant $k$, are predicted by

$$\hat{x}_{k+1|k} = \hat{x}_k + T f(\hat{x}_k, u_k). \tag{B.5}$$

The measurement prediction reads

$$\hat{y}_{k+1|k} = g(\hat{x}_{k+1|k}). \tag{B.6}$$

To obtain the covariance of the state prediction $\hat{x}_{k+1|k}$, the nonlinear representation (B.2) is linearised with respect to the true (deterministic) quantities,

$$\hat{x}_{k+1|k} \doteq x_{k+1} + \underbrace{\left(T\frac{\partial f}{\partial x}\bigg|_{x=\hat{x}_k} + I\right)}_{A_k}(\hat{x}_k - x_k) + T\frac{\partial f}{\partial u}\bigg|_{u=\hat{u}_k}\underbrace{(\hat{u}_k - u_k)}_{0} + \underbrace{T\frac{\partial f}{\partial w}\bigg|_{w=w_k}}_{I} w_k \qquad (B.7)$$

where the variation of the input $u$ vanishes, because it is assumed to be deterministic and the dependency of the system noise with respect to the states is assumed to be unity without loss of generality. The evaluation of the Jacobian $A_k$ at the estimated states $\hat{x}_k$ instead of the true states $x_k$ is an approximation out of necessity, since the true states are not known.
Then the covariance of the state prediction $\hat{x}_{k+1|k}$, defined by

$$P_{k+1|k} := E\{\hat{x}_{k+1|k}\hat{x}_{k+1|k}^T\}, \qquad (B.8)$$

where $E\{\cdot\}$ denotes the expectation operator, is given by

$$P_{k+1|k} = A_k P_k A_k^T + Q, \qquad (B.9)$$

where the covariances of the true states (without hat) in (B.7) vanish.
The update of the states is given by

$$\hat{x}_{k+1} = \hat{x}_{k+1|k} + K_k(y_{k+1} - \hat{y}_{k+1|k}), \qquad (B.10)$$

where $K_k$ denotes the Kalman gain matrix.
The linearisation of the measurement equation (B.3) with respect to the true quantities is calculated as

$$\hat{y}_{k+1|k} \doteq y_{k+1} + \underbrace{\frac{\partial g}{\partial x}\bigg|_{x=\hat{x}_{k+1|k}}}_{C_k}(\hat{x}_{k+1|k} - x_{k+1}) + \underbrace{\frac{\partial g}{\partial v}\bigg|_{v=v_k}}_{I} v_k. \qquad (B.11)$$

Again, the evaluation of the Jacobian $C_k$, also called measurement matrix, is approximately performed for the estimated states, and the derivative of the measurements with respect to the noise is assumed to be unity.
Inserting (B.11) into (B.10) yields

$$\hat{x}_{k+1} = \hat{x}_{k+1|k} - K_k C_k \hat{x}_{k+1|k} + K_k C_k x_{k+1} + K_k v_k. \qquad (B.12)$$

Application of the expectation operator on the square of the final estimate leads to its covariance,

$$P_{k+1} = (I - K_k C_k) P_{k+1|k} (I - K_k C_k)^T + K_k R K_k^T. \qquad (B.13)$$

Now the Kalman gain matrix $K_k$ is calculated by minimisation of the trace of the state covariance matrix, which is identical to the estimation error covariance matrix. Thereby the variances of the estimation errors are minimised.

The problem is formulated as

$$\text{Tr}[\boldsymbol{P}_{k+1}] \to \min_{\boldsymbol{K}_k}. \tag{B.14}$$

Using two equalities known from matrix calculus

$$\frac{\partial \text{Tr}[\boldsymbol{ACA}^\text{T}]}{\partial \boldsymbol{A}} = 2\boldsymbol{AC}, \quad \boldsymbol{C} \text{ symmetric} \tag{B.15}$$

and

$$\frac{\partial \text{Tr}[\boldsymbol{AB}^\text{T}]}{\partial \boldsymbol{A}} = \boldsymbol{B} \quad \text{or} \quad \frac{\partial \text{Tr}[\boldsymbol{BA}^\text{T}]}{\partial \boldsymbol{A}} = \boldsymbol{B}^\text{T}, \quad \boldsymbol{AB}^\text{T} \text{ square,} \tag{B.16}$$

the derivative of the trace of $\boldsymbol{P}_{k+1}$ with respect to $\boldsymbol{K}_k$ is calculated and set equal to zero,

$$\frac{\partial \text{Tr}[\boldsymbol{P}_{k+1}]}{\partial \boldsymbol{K}_k} = 2\boldsymbol{K}_k\boldsymbol{C}_k\boldsymbol{P}_{k+1|k}\boldsymbol{C}_k^\text{T} + 2\boldsymbol{K}_k\boldsymbol{R} - 2\boldsymbol{P}_{k+1|k}\boldsymbol{C}_k^\text{T} \stackrel{!}{=} 0. \tag{B.17}$$

To obtain the latter expression, the symmetry of the covariance matrix is exploited. Solving for $\boldsymbol{K}_k$ yields

$$\boldsymbol{K}_k = \boldsymbol{P}_{k+1|k}\boldsymbol{C}_k^\text{T}(\boldsymbol{C}_k\boldsymbol{P}_{k+1|k}\boldsymbol{C}_k^\text{T} + \boldsymbol{R})^{-1}. \tag{B.18}$$

Thereby the equations of the Extended Kalman Filter are complete.

## B.2 Practical issues

Sometimes the system noise cannot be obtained from physical modelling. In some cases it can then be entirely omitted, but sometimes it needs to be systematically increased until convergence is attained.

The measurement noise is found from the sensor manufacturer's data sheets, or, more reliably, from recorded steady-state signals of the sensors. It can, however, improve the performance of the filter to work with a different noise covariance than actually present. This is subject to systematic tuning.

For both types of noise the covariances are usually assumed to be constant, but in some cases it may be advisable to introduce time-dependency.

Another issue is the choice of the initial state covariance. It may influence the performance of the filter and is also found by systematic tuning.

# Bibliography

[1] A.P. Aguiar and A. Pascoal. Stabilization of the Extended Nonholonomic Double Integrator via Logic-Based Hybrid Control. In *6th IFAC Symposium on Robot Control*, Vienna, Austria, 2000.

[2] Analog Devices. ADSP BF533 High Performance General Purpose Blackfin Processor Data Sheet, August 2006. http://www.analog.com/UploadedFiles/Data_Sheets/126191093ADSPBF533_c.pdf.

[3] Analog Devices. ADXL202 ±2g Dual Axis Accelerometer Data Sheet, August 2006. http://www.analog.com/UploadedFiles/Data_Sheets/53728567227477ADXL202E_a.pdf.

[4] Analog Devices. ADXRS300 Angular Rate Sensor Data Sheet, August 2006. http://www.analog.com/UploadedFiles/Data_Sheets/732884779ADXRS300_b.pdf.

[5] Analog Devices. VisualDSP++ 3.5 DSP Development Environment, August 2006. http://www.analog.com/processors/blackfin/evaluationDevelopment/blackfinProcessorTestDrive.html.

[6] E. Anderson, Z. Bai, C. Bischof, S. Blackford, J. Demmel, J. Dongarra, J. Du Croz, A. Greenbaum, S. Hammarling, A. McKenney, and D. Sorensen. *LAPACK Users' Guide*. Society for Industrial and Applied Mathematics, Philadelphia, 1999.

[7] S. Aydin and H. Temeltas. A Novel Approach to Smooth Trajectory Planning of a Mobile Robot. In *IEEE International Conference on Advanced Motion Control*, Maribor, Slovenia, 2002.

[8] R. Balakrishna and A. Ghosal. Modeling of Slip for Wheeled Mobile Robots. *IEEE Transactions on Robotics and Automation*, 11(1), 1995.

[9] J. Barraquand, B. Langlois, and J.-C. Latombe. Numerical Potential Field Techniques for Robot Path Planning. *IEEE Transactions on Systems, Man and Cybernetics*, 22(2), 1992.

[10] B. Barshan and H.F. Durrant-Whyte. Inertial Navigation Systems for Mobile Robots. *IEEE Transactions on Robotics and Automation*, 11, 1995.

[11] Basler Vision Technologies. A301FC Basler Firewire CCD-Camera Data Sheet, August 2006. http://www.baslerweb.com/downloads/10577/DA00044803_A301f_Users_Manual.pdf.

[12] Basler Vision Technologies. BCAM 1394 Driver 1.8 for CCD-Camera, August 2006. http://www.baslerweb.com/beitraege/maildownload_formular_de_31330.html.

[13] A. Bloch and S. Drakunov. Stabilization of a Nonholonomic System via Sliding Modes. In *IEEE Conference on Decision and Control*, Lake Buena Vista (FL), USA, 1994.

[14] J. Borenstein, H.R. Everett, and L. Feng. *Where am I? Sensors and Methods for Mobile Robot Positioning*. Technical Report, University of Michigan, 1996.

[15] J. Borenstein and L. Feng. Gyrodometry: A New Method for Combining Data from Gyros and Odometry in Mobile Robots. In *IEEE International Conference on Robotics and Automation*, Minneapolis (MN), USA, 1996.

[16] J. Borenstein and L. Feng. Measurement and Correction of Systematic Odometry Errors in Mobile Robots. *IEEE Transactions on Robotics and Automation*, 12, 1996.

[17] G.A. Borges and M.-J. Aldon. Environment Mapping and Robust Localization for Mobile Robots Navigation in Indoor Environments. In *Congresso Brasileiro de Automatica*, Gramado, Brazil, 2004.

[18] M. Breivik and T.I. Fossen. Guidance-Based Path Following for Wheeled Mobile Robots. In *16th IFAC world congress*, Prague, Czech Republic, 2005.

[19] R.W. Brockett. *Differential Geometric Control Theory*, chapter Asymptotic Stability and Feedback Stabilization. Birkhauser, 1983.

[20] J. Burlet, O. Aycard, and T. Fraichard. Robust Motion Planning using Markov Decision Processes and Quadtree Decomposition. In *IEEE International Conference on Intelligent Robots and Systems*, New Orleans (LA), USA, 2004.

[21] C. Canudas de Wit, H. Khennouf, C. Samson, and O.J. Sördalen. *The Book of Mobile Robots*, chapter Nonlinear Control Design of Mobile Robots. World Scientific, Singapore, 1996.

[22] L. Caracciolo, De Luca A., and S. Iannitti. Trajectory Tracking Control of a Four-Wheeled Differentially Driven Mobile Robot. In *IEEE International Conference on Robotics and Automation*, Detroit (MI), USA, 1999.

[23] B. d'Andrea Novel, G. Campion, and G. Bastin. Control of Nonholonomic Wheeled Mobile Robots by State Feedback Linearization. *The International Journal of Robotics Research*, 14(6), 1995.

[24] A. De Luca and M. D. Di Benedetto. Control of Nonholonomic Systems Via Dynamic Compensation. *Kybernetica*, 29(6), 1993.

[25] R.A. DeCarlo, S.H. Zak, and G.P. Matthews. Variable Structure Control of Nonlinear Multivariable Systems. *Proceedings of the IEEE*, 76(3), 1988.

[26] A.W. Divelbiss and J.T. Wen. Trajectory Tracking Control of a Car-Trailer System. *IEEE Transactions on Control Systems Technology*, 5(3), 1997.

[27] L.E. Dubins. On Curves of Minimal Length with a Constraint on Average Curvature, and with Prescribed Initial and Terminal Positions and Tangents. *American Journal of Mathematics*, 79, 1957.

[28] M. Egerstedt, X. Hu, and A. Stotsky. Control of Mobile Platforms Using a Virtual Vehicle Approach. *IEEE Transactions on Automatic Control*, 46(11), 2001.

[29] A. Elnagar and A. Basu. Heuristics for Local Path Planning. *IEEE Transactions on Systems, Man and Cybernetics*, 23(2), 1993.

[30] A. Elnagar and A.M. Hussein. Acceleration-Based Optimal Trajectory Planning in 3D Environments. In *IEEE International Conference on Intelligent Robots and Systems*, Victoria (BC), Canada, 1998.

[31] T. Ersson and X. Hu. Path Planning and Navigation of Mobile Robots in Unknown Environments. In *IEEE International Conference on Intelligent Robots and Systems*, Maui (HI), USA, 2001.

[32] E. Fabrizi, G. Oriolo, S. Panzieri, and G. Ulivi. Enhanced Uncertainty Modeling for Robot Localization. In *7th International Symposium on Robotics with Application*, Anchorage (AK), USA, 1998.

[33] E. Fabrizi, G. Oriolo, S. Panzieri, and G. Ulivi. Mobile Robot Localization via Fusion of Ultrasonic and Inertial Sensor Data. In *8th International Symposium on Robotics with Application*, Maui (HI), USA, 2000.

[34] Faulhaber. 2224SR DC-Micromotor Data Sheet, August 2006. http://www.minimotor.ch/minicatalog/pdf/DC-Motors/eM2224SR.pdf.

[35] D. Ferguson and A. Stentz. The Delayed D* Algorithm for Efficient Path Replanning. In *IEEE 2005 International Conference on Robotics and Automation*, Barcelona, Spain, 2005.

[36] R. Fierro and F.L. Lewis. Control of a Nonholonomic Mobile Robot: Backstepping Kinematics into Dynamics. *Journal of Robotics Systems*, 14(3), 1997.

[37] M. Fliess, J. Levine, P. Martin, and P. Rouchon. Design of trajectory stabilizing feedback for driftless flat systems. In *European Control Conference*, Rome, Italy, 1995.

[38] T. Fraichard. Dynamic Trajectory Planning with Dynamic Constraints: a 'State-Time Space' Approach. In *IEEE International Conference on Intelligent Robots and Systems*, Yokohama, Japan, 1993.

[39] T. Fraichard and C. Laugier. Path-Velocity Decomposition Revisited and Applied to Dynamic Trajectory Planning. In *IEEE International Conference on Robotics and Automation*, Atlanta (GA), USA, 1993.

[40] T. Fraichard and A. Scheuer. From Reeds and Shepp's to Continuous-Curvature Paths. *IEEE Transactions on Robotics*, 20(6), 2004.

[41] Y. Fuke and E. Krotkov. Dead Reckoning for a Lunar Rover on Uneven Terrain. In *IEEE International Conference on Robotics and Automation*, Minneapolis (MN), USA, 1996.

[42] J.E. Guivant and E.M. Nebot. Optimization of the Simultaneous Localization and Map-Building Algorithm for Real-Time Implementation. *IEEE Transactions on Robotics and Automation*, 17(3), 2001.

[43] M. Haddad, T. Chettibi, S. Hanchi, and H.E. Lehtihet. A New Approach for Minimum Time Motion Planning Problem of Wheeled Mobile Robots. In *16th IFAC world congress*, Prague, Czech Republic, 2005.

[44] J. Hespanha, D. Liberzon, and A.S. Morse. Logic-Based Switching Control of a Nonholonomic System with Parametric Modeling Uncertainty. *Systems & Control Letters, Special Issue on Hybrid Systems*, 38(3), 1999.

[45] J.P. Hespanha and A.S. Morse. Stabilization of Nonholonomic Integrators via Logic-Based Switching. *Automatica's Special Issue on Hybrid Systems*, 35, 1999.

[46] T. Hu and S.X. Yang. Real-time Motion Control of a Nonholonomic Mobile Robot with Unknown Dynamics. In *Computational Kinematics Conference*, Seoul, Corea, 2001.

[47] A.M. Hussein and A. Elnagar. On Smooth and Safe Trajectory Planning in 2D Environments. In *IEEE International Conference on Robotics and Automation*, Albuquerque (NM), USA, 1997.

[48] A.M. Hussein and A. Elnagar. Motion Planning Using Maxwell's Equations. In *IEEE International Conference on Intelligent Robots and Systems*, Lausanne, Switzerland, 2002.

[49] Infineon. XC167CI 16bit Microcontroller Data Sheet, August 2006. http://www.infineon.com/cgi-bin/ifx/portal/ep/channelView.do?channelId==-64419&channelPage=2Fep2Fchannel2FproductOverview.jsp&pageTypeId=17099.

[50] A. Isidori. *Nonlinear Control Systems*. Springer, Wien, New York, 1989.

[51] P. Jacobs and J. Canny. Planning Smooth Paths for Mobile Robots. In *IEEE International Conference on Robotics and Automation*, Scottsdale (AZ), USA, 1989.

[52] S. Jakubek, M. Seyr, and G. Novak. Autonomous Mobile Robot Proprioceptive Navigation and Predictive Trajectory Tracking. *Robotics and Autonomous Systems, under review*, 2006.

[53] Z.-P. Jiang and H. Nijmeijer. Tracking Control of Mobile Robots: A Case Study in Backstepping. *Automatica*, 33(7), 1997.

[54] Z.-P. Jiang and H. Nijmeijer. A Recursive Technique for Tracking Control of Nonholonomic Systems in Chained Form. *IEEE Transactions on Automatic Control*, 44(2), 1999.

[55] R.E. Kalman. A New Approach to Linear Filtering and Prediction Problems. *Transactions of the ASME-Journal of Basic Engineering*, 82, 1960.

[56] Y. Kanayama and B.I. Hartman. Smooth Local Path Planning for Autonomous Vehicles. In *IEEE International Conference on Robotics and Automation*, Scottsdale (AZ), USA, 1989.

[57] Y. Kanayama, Y. Kimura, F. Miyazaki, and T. Noguchi. A Stable Tracking Control Method for an Autonomous Mobile Robot. In *IEEE International Conference on Robotics Automation*, Cincinnati (OH), USA, 1990.

[58] A. Kelly and R. Unnikrishnan. Efficient Construction of Optimal and Consistent Ladar Maps using Pose Network Topology and Nonlinear Programming. In *11th International Symposium of Robotics Research*, Siena, Italy, 2003.

[59] B. Kim and P. Tsiotras. Controllers for Unicycle-Type Wheeled Robots: Theoretical Results and Experimental Validation. *IEEE Transactions on Robotics and Automation*, 18(3), 2002.

[60] I. Kolmanovsky and N.H. McClamroch. Developments in Nonholonomic Control Problems. *IEEE Control Systems Magazine*, 15(6), 1995.

[61] J.-C. Latombe. *Robot Motion Planning*. Kluwer Academic Publishers, Boston, Dordrecht, London, 1991.

[62] J.-P. Laumond, P.E. Jacobs, M. Taix, and R.M. Murray. A Motion Planner for Nonholonomic Mobile Robots. *IEEE Transactions on Robotics and Automation*, 10(5), 1994.

[63] T.-C. Lee, K.-T. Song, C.-H. Lee, and C.-C. Teng. Tracking Control of Unicycle-Modeled Mobile Robots Using a Saturation Feedback Controller. *IEEE Transactions on Control Systems Technology*, 9(2), 2001.

[64] J. Leonard, H.F. Durrant-Whyte, and I.J. Cox. Dynamic Map Building for an Autonomous Mobile Robot. In *IEEE International Workshop on Intelligent Robots and Systems*, Tsuchiura, Ibaraki, Japan, 1990.

[65] C.-S. Liu and H. Peng. Road Friction Coefficient Estimation For Vehicle Path Prediction. *Vehicle System Dynamics*, 25, 1996.

[66] J. Luo and P. Tsiotras. Exponentially Convergent Control Laws for Nonholonomic Systems in Power Form. *Systems and Control Letters*, 35, 1998.

[67] R. Mazl and L. Preucil. Sensor Data Fusion for Inertial Navigation of Trains in GPS-dark Areas. In *IEEE Intelligent Vehicles Symposium*, Columbus (OH), USA, 2003.

[68] B. Mirtich and J. Canny. Using Skeletons for Nonholonomic Path Planning among Obstacles. In *IEEE International Conference on Robotics and Automation*, New York City (NY), USA, 1992.

[69] P.F. Muir and C.P. Neuman. Kinematic Modelling of Wheeled Mobile Robots. *Journal of Robotic Systems*, 4(2), 1987.

[70] R.M. Murray, Z. Li, and S.S. Sastry. *A Mathematical Introduction to Robotic Manipulation*. CRC Press, Boca Raton (FL), USA, 1994.

[71] M. Norgaard, O. Ravn, and N.K. Poulsen. *Neural Networks for Modelling and Control of Dynamic Systems*. Springer, Wien, New York, 1999.

[72] G. Novak and S. Mahlknecht. TINYPHOON - A Tiny Autonomous Mobile Robot. In *IEEE International Symposium on Industrial Electronics*, Dubrovnik, Croatia, 2005.

[73] K. Ogata. *Discrete Time Control Systems*. Prentice-Hall, Englewood Cliffs (NJ), USA, 1987.

[74] G. Oriolo, A. De Luca, and M. Vendittelli. WMR Control Via Dynamic Feedback Linearization: Design, Implementation, and Experimental Validation. *IEEE Transactions on Control Systems Technology*, 10(6), 2002.

[75] G. Oriolo, S. Panzieri, and G. Ulivi. Cyclic Learning Control for Chained-Form Systems with Application to the Car-Like Robot. In *13th IFAC World Congress*, San Francisco (CA), USA, 1996.

[76] H.B. Pacejka. *Tyre and vehicle dynamics*. Society of Automotive Engineers, 2002.

[77] K. Park, C. Hakyoung, and J.-G. Lee. Point Stabilization of Mobile Robots via State-Space Exact Feedback Linearization. In *IEEE International Conference on Robotics and Automation*, Detroit (MI), USA, 1999.

[78] A. Piazzi and C. Guarino Lo Bianco. Quintic $G_2$-Splines for Trajectory Planning of Autonomous Vehicles. In *IEEE Intelligent Vehicles Symposium*, Dearborn (MI), USA, 2000.

[79] J.A. Reeds and L.A. Shepp. Optimal Path for a Car That Goes Both Forwards and Backwards. *Pacific Journal of Mathematics*, 1(145), 1990.

[80] S.I. Roumeliotis and G.A. Bekey. An Extended Kalman Filter for frequent local and infrequent global sensor data fusion. In *International Symposium on Intelligent Systems and Advanced Manufacturing*, Pittsburgh (PA), USA, 1997.

[81] C. Samson. Control of Chained Systems. Application to Path Following and Time-Varying Point-Stabilization of Mobile Robots. *IEEE Transactions on Automatic Control*, 40(1), 1995.

[82] M. Seyr and S. Jakubek. Mobile robot predictive trajectory tracking. In *International Conference on Informatics in Control, Automation and Robotics*, Barcelona, Spain, 2005.

[83] M. Seyr and S. Jakubek. Proprioceptive Navigation, Slip Estimation and Slip Control for Autonomous Wheeled Mobile Robots. In *IEEE International Conference on Robotics, Automation and Mechatronics*, Bangkok, Thailand, 2006.

[84] M. Seyr and S. Jakubek. Dynamic Trajectory Generation via Numerical Multi-Objective Optimisation. In *American Control Conference, under review*, New York City (NY), USA, 2007.

[85] M. Seyr, S. Jakubek, and G. Novak. Neural Network Predictive Trajectory Tracking of an Autonomous Two-Wheeled Mobile Robot. In *16th IFAC world congress*, Prague, Czech Republic, 2005.

[86] E. Solda, R. Worst, and J. Hertzberg. Poor Man's Gyro-Based Localization. In *IFAC/EURON Symposium on Intelligent Autonomous Vehicles*, Lisboa, Portugal, 2004.

[87] A. Stentz. Optimal and Efficient Path Planning for Partially-Known Environments. In *IEEE International Conference on Robotics and Automation*, San Diego (CA), USA, 1994.

[88] STMicroelectronics. L298N Dual H-Bridge Data Sheet, August 2006. http://www.st.com/stonline/products/literature/ds/1773/l298.pdf.

[89] H.G. Tanner and K.J. Kyriakopoulos. Discontinuous Backstepping for Stabilization of Nonholonomic Mobile Robots. In *IEEE International Conference on Robotics and Automation*, Washington DC (MD), USA, 2002.

[90] The Mathworks. Matlab 7.0, Release 14, August 2006. http://www.mathworks.de/products/matlab/.

[91] S. Thrun. *Exploring Artificial Intelligence in the New Millenium*, chapter Robotic Mapping: A Survey. Morgan Kaufmann, 2002.

[92] S. Thrun, D. Fox, and W. Burgard. Probabilistic Mapping Of An Environment By A Mobile Robot. In *IEEE International Conference on Robotics and Automation*, Leuven, Belgium, 1998.

[93] P. Tsiotras. *Modeling and Control of Mechanical Systems*, chapter Invariant Manifold Techniques for Control of Underactuated Mechanical Systems. Imperial College, London, U.K., 1997.

[94] V.I. Utkin. Variable Structure Systems with Sliding Modes. *IEEE Transactions on Automatic Control*, 22(2), 1977.

[95] J. Villagra and H. Mounier. Obstacle-Avoiding Path Planning for High Velocity Wheeled Mobile Robots. In *16th IFAC world congress*, Prague, Czech Republic, 2005.

[96] VirtualDub. VirtualDub 1.5.10, August 2006. http://www.virtualdub.org.

[97] H.-J. von der Hardt, D. Wolf, and R. Husson. The Dead Reckoning Localization System of the Wheeled Mobile Robot ROMANE. In *IEEE International Conference on Multisensor Fusion and Integration for Intelligent Systems*, Washington DC (MD), USA, 1996.

[98] Y. Wang and D. Mulvaney. Genetic-based Mobile Robot Path Planning using Vertex Heuristics. In *IEEE International Conference on Cybernetics and Intelligent Systems*, Bangkok, Thailand, 2006.

[99] Wikipedia. Entry for the term autonomy, September 2006. http://en.wikipedia.org/wiki/Autonomous.

[100] R.L. Williams, B.E. Carter, P. Gallina, and G. Rosati. Dynamic Model With Slip for Wheeled Omnidirectional Robots. *IEEE Transactions on Robotics and Automation*, 18(3), 2002.

[101] W. Wu, H. Chen, and Y. Wang. Backstepping Design for Path Tracking of Mobile Robots. In *IEEE International Conference on Intelligent Robots and Systems*, Piscataway (NJ), USA, 1999.

[102] J.-M. Yang, I.-H. Choi, and J.-H. Kim. Sliding Mode Control of a Nonholonomic Wheeled Mobile Robot for Trajectory Tracking. In *IEEE International Conference on Robotics and Automation*, Leuven, Belgium, 1988.

Die VDM Verlagsservicegesellschaft sucht für wissenschaftliche Verlage abgeschlossene und herausragende

## Dissertationen, Habilitationen, Diplomarbeiten, Master Theses, Magisterarbeiten usw.

für die kostenlose Publikation als Fachbuch.

Sie verfügen über eine Arbeit, die hohen inhaltlichen und formalen Ansprüchen genügt, und haben Interesse an einer honorarvergüteten Publikation?

Dann senden Sie bitte erste Informationen über sich und Ihre Arbeit per Email an *info@vdm-vsg.de*.

**Sie erhalten kurzfristig unser Feedback!**

VDM Verlagsservicegesellschaft mbH
Dudweiler Landstr. 99        Telefon  +49 681 3720 174
D - 66123 Saarbrücken        Fax      +49 681 3720 1749
**www.vdm-vsg.de**

Die VDM Verlagsservicegesellschaft mbH vertritt

Printed by Books on Demand GmbH, Norderstedt / Germany